West Point. WOMAN

HOW CHARACTER IS CREATED AND LEADERSHIP IS LEARNED

PRAISE FOR
WEST POINT WOMAN

"The U.S. Military Academy, West Point, did not admit women until 1976. When it finally did, the decision provoked controversy. Into that atmosphere walked Sara Potecha, who started attending West Point in 1979 as a member of only the fourth co-ed class in the school's history. The experiences she had there are lessons about failure and perseverance, humility and resilience, camaraderie and trust. To some degree, they are lessons any graduate of West Point could share, but in Sara's hands, they carry special resonance and weight. During her years at the U.S. Military Academy, she not only had to shatter preconceptions, she had to learn how to thrive often despite the system, rather than because of it. Humor, encouragement and stoicism all proved invaluable during her time in the crucible that refined her character."

Thomas F. Farrell II
Chairman, President and Chief Executive Officer, Dominion Energy

"Sara's story is one of triumph over tragedy, of transcending bias to excel in her military, personal and professional lives. No matter our age, race, gender identity or experience, we all can learn from her story. With a healthy dose of humility and introspection, she shows us the power we have to determine our own futures through our attitude and approach to the obstacles that will block our path."

Lori Herndon
President and Chief Executive Officer, AtlantiCare

"Sara Potecha and I began our careers in parallel universes on two different rivers, mine along the Severn in Annapolis, and hers along the meandering Hudson at West Point. I know that neither of us set out to prove what a minority or a woman could achieve — those conditions were forced upon us. Our goal was to serve our country the best way we could. It was others who defined our potential by how we looked or which chromosomes we had, and it was that external perspective that made our circumstances different from the other people we served with. In this wonderful, timely book, Potecha has found a way to give voice to my experiences and those of hundreds of other women and minorities who chose to attend an elite service academy. Her words are the threads that sew each of our experiences together, weaving them into a tapestry of shared sacrifice, commitment, and love of country. Her story is our story, and it reflects the best of what America is all about."

Jeff Brown

USNA '84, Former Naval Aviator, Advisor to the Governor of Virginia

"*West Point Woman* is one of the best books on leadership that I have encountered. Sara Potecha takes her unique experiences from the business world, motherhood and as one of the first women graduates from West Point and blends them all into stories that teach. Potecha's style makes the stories both relatable and translatable into actions that can be taken to improve the reader's outlook on perceived challenges, obstacles and ability to lead. As a mother and a woman in the corporate world, I found this book inspired me to continue to work on my own leadership style, learn from my challenges, and continue to lead and develop those around me with both discipline and compassion."

Michele McCauley

SVP Human Resources, Apex Systems | Apex Life Sciences

"An inspiring memoir full of passion, adversity, triumph, and thought leadership. Sara Potecha provides an insightful first-person perspective of her experience as one of the first women graduates from West Point military academy. Captivating stories capture how her experiences were molded into a leadership philosophy and helped transform her into a highly sought-after corporate management consultant. Think you're having a bad day? Stuck in the mud? *West Point Woman* will help you see things through a different lens and broaden your leadership perspective along the way."

Houston Mills

Former United States Marine Corps (USMC) Officer
and F/A-18 Fighter Pilot, Airline Executive

"We learn best through stories; we grow best through adversity. Sara Potecha applies these ideas in her leadership memoir, *West Point Woman: How Character is Created and Leadership is Learned*. Professionally and personally, Potecha has experienced unique challenges and has demonstrated extraordinary resilience and grit. As one of the first women graduates at West Point, as a management consultant in the corporate world, as a young widow with three small children, Potecha shares her journey with us. In Sara's poignant yet practical, thought-provoking yet easy-to-read voice, *West Point Woman* translates her own story into leadership lessons for all."

Gail O'Sullivan Dwyer

USMA '81, Author of *Tough As Nails: One Woman's Journey through West Point*

"In *West Point Woman*, Sara describes and gives credit to nine fundamentals that helped her through her journey to success. They are the tenets that result in strength of character and reinforce fair play and teamwork, while also accepting accountability as a professional and a human being. I encourage all of us to read this book and remind ourselves of what it takes to be a winner and a valued person. Let us endeavor as women and men to share these fundamentals and work as equal partners in making our world a better place for all."

Rand Blazer
President, Apex Segment

"In *West Point Woman*, Sara (Fotsch) Potecha captures an extraordinary perspective of one of West Point's first female graduates. Her 'path-finding' story of faith, courage, resilience and humor is a must-read for anyone determined to scale lofty heights against all odds. Sara instantly connects the reader with her personality — and her 'boots' — as she walks through and beyond one of the most significant barriers of American history. Having attended both West Point and Annapolis provides me only a deeper appreciation for the 'unique' challenges she conquered in those early years!"

Chris Williams
USNA '85, Former Navy Fighter Pilot, Airline Executive

"It was my great privilege and honor to have been involved in testing the first class of women candidates to enter West Point in 1976. Little did I know that one day I would have the pleasure of hiring a West Point woman to join my company. While Sara contributed in so many ways to the success of our firm, her courage and character in the face of personal tragedy inspired us all. I was excited when reading *West Point Woman* to see how Sara modeled leadership by using her real-life experiences and principles learned at our alma mater. Sara is a winner!"

Robert Bryant
USMA '72, CFO, Glavé & Holmes Architecture

"First and foremost, it is certainly an honor to have this opportunity to stand alongside and publicly recognize my classmate, Sara Potecha. Her work captures the essence of her experience — and many of ours — at a time of historical change for an institution recognized for honored traditions, service to one's country, and dedication to the defense of our grand nation. Sara's ability to commit her experiences to written word is tremendous — with elegance and detail. This is a real-life story where the triumph of the human spirit celebrates. Well done, classmate. Godspeed.

Major General Ray Royalty
Commanding General, 84th Training Command

"In *West Point Woman*, Sara Potecha has eloquently shared the gift of her 'first woman' experiences at the Academy with us to refocus our leadership attributes, both personally and professionally. While peeling back the layers, Sara exposes the balancing act of vulnerability and oftentimes humor required to be an effective leader — one who can work through his or her own obstacles and challenges while also leading others through their own. Sara is able to personify the principle of integrity by imploring readers to lead by action and not by words."

Tierra Kavanaugh Wayne
CEO, TKT & Associates

"Part storyteller, part consultant, and part historian, Sara Potecha reaches into her life and shares with us the reality of what it was like to be one of the first women to graduate from West Point Military Academy. If you want to know what it feels like to be a plebe or a cow or a captain, read this book. Her skills and values were shaped by these experiences, and she will engage you into asking the questions that she herself had to answer about family, hardship, perseverance, and leadership."

Phoebe A. Wood
Corporate Director, CompaniesWood

"There are so many lessons I took from *West Point Woman*. But the one that resonates most is that of building camaraderie. I want to dispel the notion that when women come together, we are divided. To accomplish that, we must have respect for all and be able to deal with our differences, empathize, and empower one another. I've learned from Sara to push beyond the boundaries. Most of our limits are self-made and stem from fear. So, I say take the leap. Jump off and don't look back. Everything you dream of is within you. And even if you fail, fail forward with purpose and drive."

Brandy Alvarado

Chair for the AVIXA Women's Council and Business Development

"Sara was a management consultant on a successful veterans employment project when I first encountered her. She definitely walked the walk and talked the talk that comes alive through the leadership examples in this book. She applied these skills to create a vision of the values veterans bring to businesses and was instrumental in developing a team of partners to achieve the mission. She led with humility, camaraderie and humor. The lessons on preconceptions and addressing adversity are critically important to those of us who face daily barriers. The book is also a testimonial to the power of networks for careers and families. *West Point Woman* is an inspiring and engaging book that demonstrates how leadership skills can be learned, modeled and applied at work and life."

Joseph M. Ashley, RhD

State Vocational Rehabilitation Administrator

"Motivating, inspiring, and encouraging, *West Point Woman* provides a unique and comprehensive view of what happens behind the veil of the Academy. Reading Sara's stories reminded me of our resilience as human beings and that through persistence we can accomplish our goals. I have spent over thirty years in higher education. I am always seeking new tools to motivate faculty, staff and students to achieve their full potential, never quit, and not to settle for anything less than their God-given talent will allow. Thanks to Sara, I now possess a powerful new resource to place in my toolbox."

Dr. Van C. Wilson

Associate Vice Chancellor, Student Experience and Strategic Initiatives, Virginia's Community Colleges

"The women I graduated with from West Point in 1988 were an amazing group. I look back on their perseverance and their accomplishments and am humbled to call them classmates and friends. I am not endorsing *West Point Woman* just because I am also a West Point graduate like the author. I am endorsing this book because Sara Potecha uses her immense storytelling abilities to not only paint vivid pictures of what life was like for the first women graduates of West Point, but also because she uses those stories to teach us definitive lessons on leadership. These are stories of character that will cause every reader to look inwardly and want to get better. But she does not stop there. The author provides the reader with action steps to implement in order to 'Cultivate Character.' This is an immensely readable book for men and women, for soldiers and civilians, and for leaders and followers."

Dave Anderson

USMA '88, Best Selling Author of *Becoming a Leader of Character: Six Habits that Make or Break a Leader at Work and at Home*

"*West Point Woman* gives us a very personal and insightful window into Sara's life as a female cadet, then an Army officer (while also serving as a mother) and finally as a successful business person. These experiences are intimately bound by Sara's selected character and leadership axioms, which stand the test of time.

Sara pulls us into her life as we experience vignettes of her personal and professional challenges, and her achievements — all the while offering examples of how she has seamlessly pursued different phases of her life while applying these axioms.

This book is a 'must read' if you are a West Pointer or want to learn more about what (and how) West Point does to teach life lessons, or if you have a child who has experienced or is going to experience West Point first-hand; it will certainly give you a great deal of insight to share over several conversations. Easy to read with many personal and business anecdotes, Sara's book, without being preachy, teaches you how to meet life challenges with consistent application of leadership principles. I believe you cannot help but feel touched by her special, very personal approach."

Chris Miller
USMA '80, HBS '87, Serial Entrepreneur

"Sara is a successful survivor of the crucible of the United States Military Academy system that had been set up and maintained as a men-only system for over 175 years. Again and again, she overcame the challenges and adversities of West Point as she formed her character and leadership style to evolve into a member of the Long Gray Line that extends over 200 years. Her book describes how she successfully served in the Army and then transitioned well to a civilian style of management that she fine-tuned as she worked with numerous organizations and companies. In her book, she very effectively discusses the leadership principles and styles that will greatly assist a leader who is trying to handle multiple challenges and implement solutions and improve performance of an organization."

William G. Haneke
Vietnam Veteran, Author of *Trust Not*

"I met Sara when we both worked at Dominion Energy. She brought the discipline of West Point to a constantly changing and evolving industry. That combination made her an innovative thinker and leader. For women in highly male environments, that combination is critical for success."

Eva Harvey
Retired EVP, Dominion Energy

West Point WOMAN

HOW CHARACTER IS CREATED AND LEADERSHIP IS LEARNED

SARA POTECHA

emerge
publishing

25 24 23 22 21 20   9 8 7 6 5 4 3 2

WEST POINT WOMAN: How Character is Created and Leadership is Learned

Second Edition–Copyright ©2020 Sara Potecha

Published by:
Emerge Publishing, LLC • Tulsa, OK 74137
Phone: 888.407.4447 www.Emerge.pub

Editing by: Megan Ryan
Original Cover Design: Courtney Hudson

ISBN: 978-1-949758-86-3 Paperback (Second Edition)

BISAC:
BUS109000   BUSINESS & ECONOMICS / Women in Business
BUS071000   BUSINESS & ECONOMICS / Leadership
BIO008000   BIOGRAPHY & AUTOBIOGRAPHY / Military

Printed in the United States of America

For my daughters, Larisa, Gwenyth, and Joye

and

For my favorite West Point Cadet, James Edwin Gaba, Jr.

TABLE OF CONTENTS

FOREWORD

"So, how's your spiritual life, Sara?" This seems like a strange question to ask a first-year student in introductory psychology who is seeking assistance with study techniques to improve her performance in class. I was a bit shocked at myself for asking the question, knowing it was personal (and certainly none of my business) and beyond the bounds of the additional instruction session she had requested. It was risky, particularly for a relatively junior faculty member at one of our nation's most disciplined and tradition-bound institutions: West Point! What was I thinking?

During the first few class sessions, Sara had seemed bright and interested in the subject, and although she was somewhat reticent to talk in class, when she did participate, she had something helpful to say. But her test performance didn't seem to match her level of engagement, which is why she asked to see me after class. As I remember it now, almost forty year later, we began talking about all the challenges she was facing as a first-year cadet (Plebe); how difficult it was to concentrate on all the tasks at hand, including studying; and how frustrated she was that she was not living up to the expectation she had set for herself after a successful high school career. Quite frankly, those were typical Plebe concerns.

On that day in the fall of 1979, she seemed loaded with nervous anxiety that was obviously getting in the way of her performance in class—and maybe out of class as well. She talked, and I listened. Eventually, she shared about her high school achievements, her close family, and her religious background. In my experience, most Plebes didn't share their faith journey with their professors, particularly not early in the semester.

I figured that might be a clue to what was behind her stress, so I asked THE QUESTION.

As it turned out, that question opened the door to courageous self-reflection that, over time, helped her appreciate herself and face the challenges of cadet life with greater confidence. Sara did the hard work of understanding and adapting to the West Point culture that was difficult for all Plebes and especially hostile to women. As she gained confidence, she experienced success in and out of the classroom and developed new talents that would last a lifetime. More importantly, she came to realize that the hardships were molding her character and giving her hope for the future.

I would like to think that Sara also saw that at West Point—despite the hardships she faced, particularly as a female cadet—we cared about the development of the whole person. Success in academics and military training mattered, but growing as a leader of character was at the core of the Corps.

I was a cadet a decade before Sara. There were no women in the Corps of Cadets back then, and we could hardly have imagined that West Point would ever be coed. But it happened in 1976, the year before I returned as a junior faculty member to teach psychology. The first class of women (the Class of 1980) were sophomores when I joined the faculty, and I was blessed to get to know many of the young women who blazed the trail for others like Sara to follow.

By the time Sara arrived two years later, I sensed that there were two West Points, one for men and one for women, after having heard many stories from women cadets about their travails. I don't mean to say that there were two distinct and separate academies on the grounds of West Point—all cadets went through the same crucible. But psychologically, they were very different experiences.

The same sense of self-reflection leading to understanding that Sara used to appreciate West Point is at the heart of this book. She brings her cadet and officer experiences to life through personal stories of the challenges she faced and the lessons about leadership she learned. Her stories give a vividly candid, humble, and, at times, humorous look at what the first classes of West Point women cadets and graduates endured to serve their country as officer leaders. Most impressively, Sara mines each experience to understand leadership principles that apply in any setting. These lessons make the book more than a memoir; Sara offers practical suggestions that you can use to grow as a leader of character and help you make sense of the challenges you may face in your work setting.

Sara and I only overlapped at West Point for one year. I went on to other assignments while she was still a cadet, and our paths never crossed in the Army. I returned to West Point in 1984, the year after she graduated, and spent the remainder of my military career on the faculty, retiring in 2005 as a professor of psychology and leadership and vice dean for education. I continued to study and teach leadership and, along with many dedicated faculty members, helped establish West Point's current leader development system. We worked hard to create conditions where the abuses described in this book are the rare exception and the highlights of inspirational leadership are the norm.

As you read Sara's stories and reflect on the leadership lessons she shares, I hope you will appreciate the courage it took for Sara and so many other brave women to lead the way. Many of those I was fortunate to know as cadets have, like Sara, gone on to become accomplished professionals in every sector of society. Their stories matter, and their example serves as an inspiration to all of us.

George B. "Barney" Forsythe, Ph.D.
Brigadier General, U.S. Army (Retired)
Professor Emeritus, U.S. Military Academy

A NOTE TO
MY READERS

Like many of you, I love a good story – the underdog who comes out ahead; the ugly duckling who discovers love; the unlikely hero who overcomes all the odds and wins. And so, in this book are many stories, written to convey the lessons of my West Point experience and provide you with a leadership lattice from which to develop your own leadership skill set.

I freely admit that these are my tales to tell, taken from my perspective as one of the first women to graduate from the Academy. My perceptions may be different from other women and men who experienced West Point at that time in history. And like all good stories, some tales I record may alternatively make you angry or sad, make you laugh or cry, and it is my hope that they may ultimately encourage you to keep going or help you make a needed course correction.

I caution you, however, that this book is not about easy fixes. Rather, it is about embracing a set of principles with discipline and in so doing, surmounting obstacles, setbacks and difficulties that you confront with renewed vigor and success.

To tell my tale, I divide this book into four parts, as outlined on the following pages.

PART I – THE SETTING

In all good stories, we must understand the backdrop and the conditions within which a tale unfolds. I begin with a description of the distinguishing character or ethos of West Point. I also outline the book's historical context, detailing the events leading up to the admission of women at the Academy and the forces working against our inclusion. And for those unfamiliar with the West Point leadership and educational system, I also introduce that process and some of a cadet's vocabulary. And in reading my tales, if you get confused with some of the West Point jargon, I include a glossary of cadet vocabulary in the back of the book.

PART II – THE FUNDAMENTALS

In this section, I describe nine foundational principles imbued through my West Point odyssey. I also detail how you might incorporate these tenets into your own leadership framework.

PART III – NAVIGATING THE FORCES

In the next four chapters, I relate the leadership principles I discovered when I was navigating the forces making adapting and succeeding at West Point all the more difficult. I provide suggestions on how to navigate your own turbulence.

PART IV – DIGGING DEEPER

In this last section I share very personal stories that further refined my understanding of key leadership axioms. These next three leadership tenets require greater maturity and commitment. Here you will learn how to dig deeper, ultimately challenging you to leave your own leadership legacy.

All the stories in this leadership memoir convey a West Point leadership lesson and principle. I give examples of how I embraced that value or skill in my various roles in life, first as cadet, then as an Army officer, and as an Executive Coach, employee, consultant and even as a mother. I also provide examples of leaders who have demonstrated these characteristics.

As appropriate at the end of a chapter, I give you practical suggestions on how you can then incorporate that principle into your leadership arsenal. I refer to these applications as "Cultivating Character."

HOW TO GET THE MOST OUT OF THIS BOOK

I suggest that you first read the entire book. Then go back and re-read the chapters that resonate the most with you. I have found that those stories or principles that reverberate are worth examining more deeply. I would further suggest that you begin any self-improvement process by selecting only two to three principles to improve. Keep in mind, self-improvement efforts can feel like one step forward and two steps backward for a while. Know that it takes about sixty days to acquire any new skill. That is the typical time needed to make it a habit.

Although this book is designed for individual improvement, teams, departments and entire organizations can also benefit from reading the book and, as a group, discussing its contents and suggested applications to their specific organizational needs.

So, whether you are confronting a difficult work environment, need encouragement to move beyond a personal crisis or simply want to improve your leadership skills, this West Point Woman is delighted to share these truths with you in the pages that follow.

Additional Help

Change is difficult. If you find yourself falling back into unproductive thinking and behaviors and would like some additional help, visit my websites for other tools and resources or set up a coaching call.

www.westpointwoman.com or www.sarapotecha.com

PROLOGUE

After several years as a management consultant, I was asked to speak to a Women's Business Resource Group for a large financial firm. This group of women had just completed an eighteen month development program culminating in each participant detailing to the company's upper management their own leadership philosophy.

The company, I was told, needed a speaker who would articulate a compelling leadership philosophy. My name was offered as a potential speaker along with other highly qualified orators. Later, I was told that I was selected over other candidates because I possessed a diverse and challenging background: I was one the first women to graduate from West Point.

Over the years, I had written and spoken on a variety of topics and veteran-related issues, but never on the West Point experience and how it had translated into my leadership philosophy. This would be stimulating new work, I thought, as I busied myself with developing content for the presentation.

As I created the presentation, a rush of stories came back from memory. How was I to describe my leadership philosophy without describing the nerve-wracking events of "R-Day" and "Beast Barracks"? How could I not mention the trip in uniform through Grand Central Station? Or the harrowing experience of the "Bob and Travel" or the "Dear Jane letter"? How could I not describe the overly hostile environment toward women cadets, and how could I not tell them about Eddie?

It was then that I realized the personal experiences, whether embarrassing or exhilarating, failure or formidable, of love or loss, were part and parcel of my experience. They could not be separated from the lessons learned. They were the essence of the experience. I persevered and completed the presentation. I delivered it with care. What surprised me most was the overwhelmingly positive response of the women in attendance. One woman sitting closest to me exclaimed, "Wow, I mean, wow!"

The personal experiences, whether embarrassing or exhilarating, failure or formidable, of love or loss, were part and parcel of my experience. They could not be separated from the lessons learned. They were the essence of the experience.

As I answered further questions and took in their compliments, I remembered what my dad had said when he read that the United States service academies had begun to accept women:

"Sara, it will be a unique experience for a woman."

When I was being berated by upperclassmen for not shining my shoes correctly or not knowing the reams of knowledge required, or when my hair was butchered by a barber and I was pushed physically, intellectually and emotionally beyond my endurance, I would sarcastically think, "Yeah, Dad, this is *unique* all right."

Yet, at that moment surrounded by those women leaders, I said to myself, "Dad, you were absolutely correct! It truly was a unique experience and it made me what I am today. Thank you for encouraging me to go to West Point and to never, ever give up."

I lived within a system that was not welcoming to women cadets. This was made evident through many in the officer ranks of the Academy staff. And harassment was all too prevalent within the cadet-run Corps of Cadets. Women at West Point, like all firsts, had to work harder to be accepted. We were ostracized and harassed simply because we were running up against tradition and prejudice.

In what could be an unforgiving and hostile environment for a woman, I learned about perseverance. In a highly competitive and academically challenging institution, I learned failure and humility while I also developed my intellect. In an environment that demanded physical agility and strength, I learned I could do far more with my 5'2" frame than I thought possible. In an institution that taught integrity and demanded it from its members, I learned that trust was foundational to leading others. I developed resilience and learned to laugh and not take myself or my situation too seriously. And even as women were being integrated into what had been an all-male bastion, I learned the value of camaraderie and sacrificial service.

At West Point, I cried, I laughed, I failed, I despaired, I succeeded, I got hurt, I loved, I began to lead, and I overcame again and again. The entire experience shaped my thinking and formed my character. And the impact did not end at graduation. Now I too was part of the "Long Gray Line," one that extended back two centuries and continues to this day.

At West Point, I cried, I laughed, I failed, I despaired,
I succeeded, I got hurt, I loved, I began to lead, and
I overcame again and again.

After leaving the leadership conference and making my way to my flight home, I thought of all the connections I had made by simply being a West Point alumna. Both professional and personal relationships existed from simply being a graduate. Men and women graduates invested in me and vice versa, extending the impact of the common experience at our "rock-bound highland home." Perhaps it was time for me to extend what I may have gained to others, especially the next generation.

As the plane lifted off the ground, my thoughts then turned to my three daughters. Now all in their twenties, even they had not heard some of the stories I shared earlier that day. And now entering their own professional lives, why had I not shared more with them, of the wisdom I gained through the challenges and painful experiences at the United States Military Academy?

My mind then turned to my many clients and colleagues. I rarely spoke about the experiences that shaped my approaches to problems and my recommendations, concepts that I had learned when I was so very young in the cauldron of the West Point experience.

I reflected on the various challenges my clients and colleagues (both the young and the more seasoned) face working in difficult work environments, demanding roles or sometimes seemingly no-win scenarios. The axioms that I had learned as a young cadet would serve them well in their situations too. And, I thought, how many young people today would benefit from knowing some of these life lessons? I became convinced that I needed to record my West Point experience and the guiding principles that had shaped my life, not just for my progeny but for the masses.

So, I began to write and document my stories and the lessons learned.

By telling a series of stories as a "First" at West Point, I could lift the curtain on the mystique of the Academy and its integration of women. I could teach others to not only survive but thrive in similar problematic and overwhelming situations. Then those who might never attend West Point might glean the leadership principles and skills needed to address their own life "battles."

For those of you who have ever felt overwhelmed or marginalized, disrespected or dismissed, this book was written for you! When what you have tried to accomplish seems riddled with difficulty, setback and disappointment, or every effort to break through the glass ceiling has left you disappointed, this book is designed to give you the tools to move forward.

You will learn, as I have, that fortitude is found in the cauldron of challenge and defeat. And success is built upon a series of failed attempts. With the correct principles at work, you can overcome and accomplish much more than you can ever imagine.

Fortitude is found in the cauldron of challenge and defeat.

PART I
The Setting

Like any good read, we need to understand the backdrop and
the conditions from which a story unfolds. In these pages,
I begin with a description of the distinguishing character or
ethos of West Point. I also outline the book's historical context,
detailing the events leading up to the admission of women at
the Academy and the forces working against our inclusion.
And for those unfamiliar with the West Point leadership and
educational system, I also introduce that process and some
of a cadet's vocabulary. And in reading my tales, if you get
confused with some of the West Point jargon, turn to the back
pages of this book for a glossary of cadet vocabulary.

THE WEST POINT
CHARACTER

Determined to share my West Point experiences with the goal of helping others who might also be facing obstacles of their own, I began this book by searching for a word to describe the West Point culture and character. I came upon the word ethos. It is a word derived from Greek, meaning:

> *The distinguishing character, sentiment, moral nature, or guiding beliefs of a person, group, or institution; also, ethic.*[1]

Ethos is also used in sociology to describe the primary character or essence of a culture. That is the "spirit" or "chi" that informs the beliefs, customs and practices of a group or society.

From ethos, we also derive the word ethnology, which refers to the comparative and analytical study of various cultures. Ethnographers study cultures employing a variety of methods. One technique is referred to as an *emic* perspective or that of an "insider's point of view." The insider, it is believed, can better derive meanings within the culture and then develop constructs that aptly describe the culture to others. This insider's perspective then provides greater context than a purely outside analysis might uncover.

---

1   Definition of ETHOS. (n.d.). Retrieved February 22, 2018, from
    https://www.merriam-webster.com/dictionary/ethos.

A view of the main buildings and commons of the U.S. Military Academy at West Point, New York. Photo credit: Buddy Mays / Alamy Stock Photo

Given my personal experience as one of the first women to graduate, I offer an insider's view of the West Point spirit. However, in so doing, I will freely admit that my experience may vary somewhat from other women, and certainly the men, who attended the same institution.

> *Given my personal experience as one of the first women to graduate, I offer an insider's view of the West Point spirit.*

For one, my experience was a point in time. I attended from 1979 to 1983. The Class of 1983 was the fourth class of women to graduate from West Point. The first class of women were our Seniors, or "Firsties" in cadet terminology. My class was still one of the earlier classes of women cadets. Ongoing hostility from cadets, the Officer Corps and even from some of the faculty still existed. Sadly, at times, the upper-class women could be very hard on us or appear disinterested because of their own fear of reprisal from male cadets.

The experience of a cadet admitted in the early years of women at West Point might vary based upon the culture of his or her cadet company and regiment. It could be unique based upon the opportunities taken or not taken. Of the many friends, classmates and teammates I have spoken to over the years, we all agreed that we each faced hurdles we needed to overcome while attending the United States Military Academy (USMA). All cadets to some degree experienced harassment. And because we were some of the first women, many of us experienced sexual harassment, especially as defined by today's standards. Some of the women experienced more blatant harassment than my experience, others less so. None of our experiences were identical, yet the essential tenets of the culture were taught and have been preserved over the 218 years of the Academy's existence.

The West Point ethos or character, at its core, is a set of principles to live by. They are values, practices and skills instilled in graduates designed to prepare us to become effective Army officers in service to our nation.

The West Point ethos or character, at its core, is a set of principles to live by. They are values, practices and skills instilled in graduates designed to prepare us to become effective Army officers in service to our nation. Indeed, the leadership skills imbued through the West Point experience and through successive generations of West Pointers has proven so successful that businesses and non-military organizations have adopted many of those tenets in both the United States and across the world.

Yet, as one of the first women, we experienced greater scrutiny, animosity and loneliness than our male counterparts. We felt the stigma of being a first from members of the officer cadre, and from some of our family, friends and even society. We were breaking preconceptions and coming up against ignorance, prejudice and even jealousy while living in a 200-year-old institution. The first women of West Point were given the choice, then, to overcome or be overcome by a system that in many ways was designed to restrict us.

Therefore, a woman's experience created even more opportunity to learn additional skills or to learn some skills more deeply in order to persevere. It is said, for example, that one does not learn resilience without having to confront a series of obstacles. One may not learn how to succeed unless given many opportunities to fail. And one may not understand how to leverage humor to combat loneliness and isolation until confronted with those circumstances. These skills may not have been as deeply needed for our male counterparts. These were lessons that, despite the Academy, became part of our personage. At least they did for this former cadet.

The Unites States Military Academy is by default a learning organization. And as such, the West Point ethos is imparted to current cadets and subsequent classes through educational training and real time learning experiences. And so I believe that these principles can be taught or for the purposes of this book, "caught" (i.e. communicated through stories).

Regardless, without needing to graduate from this venerable institution, you can apply the West Point tenets to your particular situation. I believe that, when consistently employed, these principles lead to success and satisfaction not just in your career, but in life. The West Point ethos can be the foundation upon which you can meet the demands and challenges for all of life's circumstances.

THE EVOLUTION OF THE WEST POINT CHARACTER

The fundamental character of USMA informs the beliefs, customs and practices of those who attend. As all such service academy institutions, West Point's ethos has evolved over time and has survived for more than two centuries because of its ability to evolve.

In fact, one could argue that the Academy has been at the forefront of the larger American cultural evolution. When President Truman called for the integration of African Americans, the Academy had to begin accepting African American men. Then, soon after the passing of the Equal Rights Amendment in 1972, a bill was introduced in Congress in 1973 directing the admission of women to all the service academies. Congressional hearings were held from May through July of 1974.

The secretaries of the Army, Air Force and Navy each testified against the admission of women. Joining them were the superintendents of each of the academies, the Department of Defense general counsel, deputy assistant secretary of defense for military personnel, the vice chief of Naval Operations, the Air Force chief of staff and the Army vice chief of staff to name a few.[2] The depth of resistance to our inclusion permeated all levels of leadership in all branches of the military.

Secretary of the Army Howard "Bo" Callaway, USMA 1949, dominated the testimony. He contended that the institution existed to produce combat arms officers.[3] Women, at the time, were barred from specific branches of the Army such as infantry, armor and special operations forces. The secretary argued that if women were admitted to the

---

2   Janda, L. (2002). *Stronger than Custom: West Point and the Admission of Women*. Westport, CT: Praeger, p.18.

3   Ibid, p.16.

Academy, West Point could not produce the number of skilled officers required for combat leader roles. What was missing from his statement was the fact that 162 graduates of the West Point Class of 1973 had selected non-combat branches of the Army, such as military intelligence, military police, quartermaster, etc.[4]

Yet, after some congressional maneuvering, the bill made it to the desk of President Ford, who signed into law the requirement that the Academies (Army, Navy, Air Force and Coast Guard) would begin to accept women starting in May 1975. Once the law was passed, the leadership of West Point had a mere eight months to plan and execute this monumental change. Not surprisingly, there was backlash from graduates, current staff and professors, and most certainly from the male cadet population.

WOMEN AT WEST POINT

One of the initial decisions made by Academy leadership was to change both the physical entrance requirements and physical training (PT) standards for women. Owing to the differences in a woman's physiology, the leaders believed these changes would aid in the recruitment of the needed number of women to fill the first class. For example, instead of pull-ups, women would perform a flexed arm hang (that is, holding themselves with their chin above the pull-up bar for as long as possible).

This decision alone created deep resentment from male cadets who felt women had it easier not only in being accepted into the Academy but in performing physically as cadets. Former graduates decried this move. "The Corps has" became the outcry of male cadets. The statement implied the Academy was in decline, that it had passed it best days. Lowering the physical standards required of women cadets deteriorated the spartan standards of the Academy. Ironically, old grads like myself

---

4   Ibid, p.20.

Female cadet demonstrating the flexed arm hang;
Source: U.S. Military Academy Library Archives Collection

now use this phrase when any major change happens at West Point, such as when cadets could opt to attend summer school to make the academic year more manageable. Obviously, that made their time at West Point less stressful than when I attended, hence "The Corps has."

Subsequently, the first women cadets were maligned, targeted and given a great deal more "attention" (hazing) from upperclassmen and staff. It was easy to see why: while the men made up 90% of the 4000 member Corps, women represented only 10%. Women could be easily identified and selected for more attention at the slightest infraction. There were many male cadets who told the first women that their sole purpose in life was to run all the women cadets out of the Corps. And many women did leave in the first classes of women. The attrition rates for women were about 50% over the first four years, while the rates for men were 30-35%.

There also was no effort by leadership to help create support *between* the women. Upper-class women were, at times, tougher on women cadets to ensure that they themselves were not perceived as showing favoritism. Like many who exist in a marginalized minority, those who survived and found acceptance by their male counterparts might distance themselves from women who struggled to avoid being grouped with them. Even with all these forces working against our assimilation, causing us additional duress and a sense of loneliness, the West Point experience was developing in women cadets what I like to call "resilient muscle." We learned survival and perseverance skills that last a lifetime.

To be clear, beyond the academics and drills, all cadets were absorbing to some degree some deeper learnings at the Academy and we may not even have been something we were aware of at the time. Nonetheless, the fundamentals of duty, honor and country became real to us. And for me, I was never the same from that point forward.

USMA MISSION

To fully understand how West Point creates leaders of character, let us begin then with the mission of the institution, as it summarizes the overall objective of the Academy:

> To educate, train, and inspire the Corps of Cadets so that each graduate is a commissioned leader of character committed to the values of *Duty, Honor, Country*, and prepared for a career of professional excellence and service to the Nation as an officer in the United States Army.[5]

Fundamentally, West Point seeks to fashion the character of each cadet so that we might serve our country with honor and excellence. This is

---

5   About West Point – Mission. (n.d.). Retrieved March 15, 2018, from https://www.usma.edu/about/SitePages/Mission.aspx.

a tall order, one that requires a great deal from the leadership and the cadet. The preparation of a cadet begins with our induction into the Fourth Class system. Initially, as New Cadets, we had to earn the right to become part of the Corps by making it through an intensive seven weeks of Cadet Basic Training (or, in cadet slang, "Beast Barracks" or "Beast"). After successfully completing Beast, we became Fourth Classmen. The upperclassmen also referred to us as Plebes. "Plebe" is borrowed from ancient Rome and is a term referring to the common people. We were also called "Beanheads" or "Smackheads." All of these titles were designed to remind us that we were to think of ourselves as the lowest of the low. Creatures who must earn our rank. As we successfully passed each year of military, physical and academic challenges, we earned greater amounts of responsibility and freedom. Third Classmen (or sophomores) were referred to in cadet vernacular as "Yearlings" or "Yuks." Then, as Second Classmen, we were called "Cows," and finally, as First Classmen, we were "Firsties."

THE FOURTH CLASS SYSTEM

Beginning as a New Cadet and throughout the fourth-class year, the Academy took away most of our civilian freedoms. Our civilian clothes were stored away. We could not have a radio or stereo in our rooms. We could not call home unless we earned the right to do so. We would not be allowed to go home until winter break. We were issued several uniforms and taught how to wear them and were reprimanded if we did not wear them perfectly. We were required to memorize a great deal of trivia, called "Plebe poop," and were expected to spew it out upon demand by the cadre. We completed runs, grass drills and road marches. We climbed ropes, swam across bodies of water carrying rifles, rappelled down mountains and learned hand-to-hand combat. We learned military drill, stood at attention for hours, and performed numerous parades. We were required as Plebes to move quickly and purposely at 120 steps per minute (referred to as "pinging").

All Fourth Classmen experienced hazing. Although no single definition of the term exists, during my tenure this could be having one or more upperclassmen stand around me as I braced my neck back and stood at attention while they used derogatory statements about my failure to wear my uniform correctly or my inability to recite Plebe poop without error. It also took its form in intensive physical exertion. "Drop and do fifty!" This demanded that I drop into the front leaning press (plank) position and complete fifty push-ups as punishment for an infraction. Extreme hazing, where a cadet's life might be endangered, was no longer allowed. Nevertheless, during my time at the Academy, hazing was emotionally and physically draining.

As Plebes, we had Fourth Class duties that often seemed demeaning, such as delivering an upperclassman's laundry or mail or memorizing an upperclassman's beverage preferences. When the academic year came

Upperclassman hazing the cadets in his squad.
Photo courtesy of Jan Tiede Swicord, USMA '83

around, we had the additional stress of completing highly competitive academic classes while performing all our military duties. And for varsity athletes like me, we trained for and competed in events, often requiring us to leave the Academy for competitions. This meant we missed classes and had to learn how to make up for lost instruction time. Cadets were forced to manage multiple demands at one time and to work with classmates to complete our duties, even knowing when to ask our classmates for help understanding a difficult subject that was eluding us. And we gladly returned the favor to our fellow cadets. We also attended honor classes to ensure we learned to make wise, honest decisions in ambiguous circumstances and that we understood and followed the Cadet Honor Code.

The Academy also insisted that all cadets were athletes, so participation on an athletic team was required. A variety of intramural sports were available to all cadets. Others would join a club squad sport or earn a spot on a varsity athletic team. Drill and ceremony were also a large component of the West Point experience, and we typically conducted at least one parade a week.

AFTER PLEBE YEAR

As upperclassmen, we were given successively more demanding leadership opportunities within the Corps. We might be a squad leader, a platoon sergeant or leader, or a company commander in a cadet company. "Striper dogs" were typically those cadets who earned top Corps leadership roles on Regimental or Brigade staff. They were identified by the many stripes on their uniforms. During our summers, we typically had a summer "detail" either leading new cadets through Cadet Basic Training or leading Yearlings through Cadet Field Training at Camp Buckner.

During a break in the summer, we might also attend a U.S. Army school, such as Airborne, Air Assault or Northern Warfare. We might be given the opportunity to serve with an Army unit at some location across the world. Other options might include shadowing an officer at the Pentagon or traveling to Africa representing the Academy. If we failed an academic or physical education class, we could attend summer school. And if you failed too many classes, you would be sent home.

INSIDER'S POINT OF VIEW

This brief overview I've shared represents a microcosm of the typical, daily USMA cadet experience. In the pages that follow, I'll share my insider's point of view on how those principles were imbued through my West Point story and have made their way into every facet of my life.

PART II
The Fundamentals

In this section, through a series of stories, I describe nine foundational principles imbued through both the triumphs and the difficulties of my West Point odyssey. These universal tenets, when consistently applied, develop leadership maturity and will lead to both personal and professional success.

CAMARADERIE

From the moment I arrived for "R-Day" (or Reception Day) as a New Cadet, I became part of a team. I was assigned to a squad, a platoon and a company. The concept of relying on our classmates was embedded in our training from day one, and it was how we were ultimately able to succeed.

Initially, we learned to work with our roommate. Our roommate helped with everything from assisting us to memorize endless fourth class knowledge, keeping the room in accordance with regulations and checking off our uniforms prior to leaving the room. By working together, we might avoid undue "attention" from upperclassmen.

Cadets don several different uniforms at the Academy, and attention to detail was paramount. As New Cadets, we learned how to polish belt buckles and shoes. Often a squad member or roommate would have mastered this skill better than others and would help their classmates with polishing or aligning the uniform correctly. The most unusual requirement, however, was the "dress-off." With any uniform that entails wearing a shirt tucked into pants or even shorts, a Fourth Classman is required to have a dress off, a body-hugging dress-off. This means that our shirts had to fit to the shape of our bodies, tailored and not loose.

A dress-off is accomplished by first unbuckling and unzipping your pants a bit. My roommate would grab the excess material on each side of my shirt, pull it back and fold it across my back. Next, I would quickly fasten my pants and belt to hold the dress-off in place. Once we moved, the dress-off would loosen of course, and we would need another dress-off.

So, it was not uncommon to ask your roommate to give you a dress-off several times a day. Women certainly have more curves than men, so our dress-offs often were found to be unacceptable despite our best efforts.

As we would conduct our Fourth Class duties, such as delivering mail or laundry to upperclassmen's rooms, our shirts would inevitably need an adjustment. Between deliveries, we might dive into a squad mate's room and ask for a dress-off to avoid being stopped and hazed. As women cadets we were constantly unzipping our pants, and our classmates (both male and female) would give us dress-offs. And then we would do the same for them. To be successful, we had to work together to complete our duties in efficient bursts of effort.

To be successful, we had to work together to complete our duties in efficient bursts of effort.

As competition was a component of every task and event at the Academy, we supported our platoon or company when competing with others. We also encouraged classmates by assisting them in preparing for physical aptitude tests, academic challenges or a military exercise. A common phrase we heard was "cooperate and graduate."

As we became upperclassmen, we worked together in leadership roles and continued to learn personal responsibility and teamwork. We began to inculcate this thinking in our young charges. When I was a Second Classman (or "Cow") and a Beast Squad leader, I had a New Cadet who came to formation still wearing his white athletic socks with his dinner uniform. The white socks were easy to see next to his gray trousers as he "pinged" by. I stopped him and his roommate. I asked the roommate, "Did you check your roommate's uniform before coming to dinner formation? "No ma'am," he said. So, I sent them both back their room to make the needed correction. The idea of camaraderie, "being in this together," had to be ingrained.

Nothing describes this esprit de corps more than when a cadet's life was cut short and he or she passed away. Our "rock bound highland home" can only be entered by driving through some steep and winding Hudson Highlands passes. I recall a classmate who was returning from a weekend pass and careened off into the deep, foreboding gullies below. After September 11, 2001, recent graduates less than a year from graduation might be serving on the front lines in the Middle East and several died in combat.

When a cadet or recent graduate passed away, just before the end of the day, the entire 4,000-member Corps of Cadets would turn off the lights in our rooms and quietly assemble outside our barracks in the darkness. We would stand at attention while Taps was played, honoring our fallen member. This ancient ritual was both moving and sobering. It made us feel connected to the deceased and to one another.

Similarly, singing in unison, "The Corps" connected us to all West Pointers who had gone before us. First learned as plebes, this hymn is frequently sung at the Academy and then afterwards at annual Founder's Day events across the world, traditionally held on or near March 16th to coincide with the date the Academy was founded in 1802. Founder's Day brings together graduates, supporters and the greater West Point family.

Camaraderie comes from our shared experience at West Point —
one of triumph and misery. As members of the Long Gray line,
we are bound together.

The Corps

The Corps, bareheaded, salute it, with eyes up, thanking our God.

That we of the Corps are treading, where they of the Corps have trod.

They are here in ghostly assemblage. The men of the Corps long dead.

And our hearts are standing attention, while we wait for their passing tread.

We Sons of today, we salute you. You Sons of an earlier day;

We follow, close order, behind you, where you have pointed the way;

The long gray line of us stretches, thro' the years of a century told

And the last man feels to his marrow, the grip of your far off hold.

Grip hands with us now though we see not, grip hands with us strengthen our hearts.

As the long line stiffens and straightens with the thrill that your presence imparts.

Grip hands tho' it be from the shadows. While we swear, as you did of yore.

Or living, or dying, to honor, the Corps, and the Corps, and the Corps.

Bishop H.S. Shipman
Former Chaplin, USMA

In 2008, the USMA Superintendent, LTG Franklin L. Hagenbeck ordered a change to the lyrics to remove gender-specific language.

Those changes include:

1.	*FROM:*	*"The men of the Corps long dead"*
	TO:	*"The ranks of the Corps long dead"*
2.	*FROM:*	*"We sons of today, we salute you"*
	TO:	*"The Corps of today, we salute you"*
3.	*FROM:*	*"You sons of an earlier day"*
	TO:	*"The Corps of an earlier day"*
4.	*FROM:*	*"And the last man feels to his marrow"*
	TO:	*"And the last one feels to the marrow"*

Camaraderie comes from our shared experience at West Point — one of triumph and misery. As members of the Long Gray Line, we are bound together. And in this hymn, the words to describe this connection are simply, "Grip hands."

Hence, as cadets we were forced to put aside differences and focus on working together for the overall benefit of our unit. We learned to appreciate the benefits of teamwork, and as leaders we knew to build this into the Army units that we would go on to command.

As a young lieutenant, I served in the 8th Infantry Division (ID) in Europe. Although disbanded now, the motto of the division was "These are my credentials." The phrase referred to the World War II event when the German unit led by Lieutenant General Ramcke surrendered to the commander of the 8th ID. Ramcke, noting that the American commanding general was lower in rank, demanded that the American verify his credentials. Brigadier General (BG) Charles D. W. Canham turned and pointed to the dirty, exhausted American GIs standing near

The author as a 1ˢᵗ Lieutenant in the 8ᵗʰ ID

him and stated, "*These* are my credentials." In that one statement, the American general acknowledged that the true victory rested not with him but with his soldiers.[1] BG Canham understood the concept of teamwork over self-promotion. It is a principle he learned while attending my alma mater and graduating in 1926.

When I was interviewing for a job a few years after leaving the service, the president of the company glanced at my resume, looked at me and said, "So, West Point and the Army? You know, you can't just give orders to employees." I smiled and reflected on the all-night maneuvers of my unit in Germany. I recalled how I gave my soldiers a mission, and even being short-handed and having equipment malfunctions, we accomplished our unit goals together. I replied, saying, "It was never about giving orders. As their leader I had to instill in them a commitment and passion to the mission of our unit. They needed to know we needed each of them to do their jobs to succeed as unit. That's how we got things done." The CEO seemed a bit taken aback by my reply. He sat for a moment and then said, "How old are you anyway?" Perhaps he thought I was older than I looked, or perhaps the CEO was skeptical that a petite woman could motivate a 300 plus organization to accomplish the mission I described. However, like many veterans, especially women veterans, I had encountered a hiring manager who was ill-equipped to interview me. One who did not understand how military leaders, to be successful, did not just give orders, but had to teach their charges to honor team over individual success.

I have worked in and with both large and small companies, all with a variety of organizational structures. Some were highly structured with clearly defined reporting structures, and others were matrixed or cross-functional with reporting structures not clearly delineated. Despite

---

1   Lone Sentry (n.d.) Retrieved March 17, 2018, from http://www.lonesentry. com/gi_stories_booklets/8thinfantry/index.html.

the type of organization, there was a common missing element of camaraderie within the workforce. Things like frequent reorganizations or
poorly planned "team-building" activities left one feeling more isolated
than connected.

Often, new leaders are selected for a management role based on their
technical expertise and not necessarily for their leadership abilities. Once
selected, many organizations do not invest in training their managers
in leadership skills such as teamwork and collaboration. It is not
surprising then that we often encounter siloed organizations and disen-
chanted workers.

Once, I was a participant in a day-long team-building event within
a large department of a Fortune 100 company. One exercise had us
divided into groups of ten. Each group occupied one table. Each participant was given an envelope with some colored shapes. We were not
allowed to speak to one another and the facilitator let us struggle initially
to determine the purpose of the shapes.

At my table sat a senior manager, a vice president in fact, "Andy." Andy
surveyed his pieces and quickly assembled them to form a rectangle.
He then folded his arms and sat back. The rest of us did not find that
our pieces formed a common geometric shape. The facilitator started
repeating team-building phrases such as, "We all do better when we use
one another" and "The best teams use everyone's contributions." Initially,
we saw no connection between what he was espousing and our task.

I began to look at my team's shapes and thought, maybe I should share
one of my pieces with someone else? So, I gave a shape away. Then a man
to my right gave a figure to the woman seated across from him. Next, she
gave him a shape in return. In time, one member completed his puzzle.
The VP, however, remained with arms crossed and did not engage with
us even though the entire team had not yet completed our figures. Other

tables were more successful than ours and were cheering after organizing all their shapes into common figures. At long last, the VP gave one of his pieces to someone, and then someone else supplied Andy with a different shape and soon we had completed the exercise successfully.

The debrief was, of course, where all the learning occurred. We were told that to be successful as a department, we needed to be willing to make the team's success more important than our personal success. The VP in question managed much like he played the game. As long as his department was getting done what needed to be done, completing their services and projects, then Andy did not feel the need to assist other departments. Camaraderie was not something he understood or valued.

We needed to be willing to make the team's success more important than our personal success.

Patrick Lencioni in his classic book, *The Advantage*, describes an effective leadership team as follows:

> Members of the leadership team are focused on team number one. They put the collective priorities and needs of the larger organization ahead of their own departments.[2]

Great teams embrace a "collective mentality" and have shared goals that overshadow individual department or personal goals. In fact, when this happens, departments will share resources rather than compete for them. Team members are willing to sacrifice for the better of the unit.[3]

---

2   Lencioni, P. (2012). *The Advantage: Why Organizational Health Trumps Everything Else in Business*, p. 71.

3   Ibid.

Army-Navy game, 2015

We learned this concept early on as New Cadets and then as Plebes helping our fellow Fourth Classmen. And then as a member of the Corps of Cadets, we stood in unison during Army football games to cheer on our usually beleaguered team. In unison, we would bellow, "The 12th man is here, the 12th man is here!" Win or lose, our chant indicated that we were an extension of our team and fully committed to it. (Unfortunately, when I attended, the Army football team lost three out of four contests with our arch rival, the Naval Academy, and tied them once). After our four years at the Academy, many of us took with us this principle of team over individual accomplishment.

Many years after graduating, I received a phone call from my classmate Mike Olsen. He called to let me know that Chip Armstrong, another USMA '83 classmate was recovering from spinal surgery at Hunter Holmes McGuire VA Medical Center in Richmond, Virginia. He suggested that we meet and visit Chip and take him to dinner. We agreed this could do a great deal to encourage him as he rehabilitated. At the time, I was living in Richmond and was only too happy to join Mike and Chip. As I spoke to Mike, I mentioned that I would be bringing my boyfriend, Reed. Also, I let Mike know that Reed was a former Navy pilot. I quickly added that he was not, however, a United States Naval Academy graduate. Mike's response intimated he was already a bit suspicious of a Navy guy, but I brushed it off and thought, Reed will enjoy meeting these guys.

When we met Mike and Chip, I had forgotten that they were both well over six feet tall, 6'3" and 6'7" in fact. My boyfriend was a mere 5'9." Reed has since commented, "Why didn't you warn me?" In my defense, when you are 5' 2", everyone is taller than you. Needless to say, as the evening progressed, my classmates began a steady stream of questions directed at Reed. What was his call sign? What did it mean? What did he fly in the Navy and what did he fly now?

As much as I enjoyed seeing my soon-to-be fiancé "sweat" during the interrogation, by night's end Reed had won them over. Mike and Chip gave me a thumbs up – demonstrating that the Navy guy had made the cut. My classmates were looking out for me; they were still living camaraderie, just like a West Pointer would do.

Camaraderie, teamwork, esprit de corps is all about looking out for one another while also holding one another accountable.

Camaraderie, teamwork, esprit de corps is all about looking out for one another while also holding one another accountable.

CULTIVATING CHARACTER

Developing camaraderie requires intention. Team members must learn to subjugate their overall ambitions to that of the group's goals. Only then can there exist a commitment to shared goals.

Yet, many of us work in environments that may not promote teamwork. Instead, we encounter toxic behaviors of co-workers: We experience back biting, purposely undermining others, gossip, agreement in meetings but no follow through, etc. We may even have team members who appear to enjoy creating unnecessary drama. How might a team member in this situation build collaboration?

Here are some suggestions to bring camaraderie into this situation:

- Model collaborative behaviors. West Point taught us to lead others by demonstrating the behaviors we wanted to see in our subordinates. This requires us to resist the urge to get in the gutter with others who exhibit negative attitudes and behaviors.

- Challenge with care the individual who may be exhibiting negative behaviors or undermining a healthy workplace. Below is a simple formula for how to confront a dysfunctional member.
    - Begin with a compliment of some positive attribute or contribution the individual has made.
    - Next, describe the negative behavior (action or words) and how it makes you feel. Avoid personal attacks. Instead make this about behavior and not about the person in general.
    - And then be quiet and listen. Sometimes people are completely unaware of how they come across.

- To further make the person comfortable, ask for feedback on your own behavior. This models humility and can even build cohesion.

- If the situation only worsens, set up a meeting with your manager to ask him or her to establish team behaviors and expectations.

- If after meeting with your manager the situation continues to deteriorate, ask to speak to an HR representative to determine other options. An experienced HR staff can often mitigate a hostile workplace culture. This option may seem counter to camaraderie. However, your mental and emotional health (and that of your other team members) need not pay the price.

All members of the team must be held to the same standards of behavior. As a leader, make the time to observe your team dynamic and address the negative behaviors of members who undermine cohesion. This means that even a teammate with a specialized skill or who is a high producer will be held responsible for disruptive, disrespectful or non-productive team behaviors.

As described above, a leader should confront a dysfunctional teammate with care. And if the negative behavior continues, the teammate should be disciplined and possibly terminated. In the end, the leader is responsible for ensuring a team is functioning optimally. Leaders can exert tremendous influence on a team by simply modeling camaraderie. That is supporting a superior or a colleague when asked. It also means the leader must avoid the appearance of favoritism within the team. A leader must avoid gossip, grudges and snap decisions. Instead a leader observes, listens and responds thoughtfully.

As I learned as a cadet, when you consistently live camaraderie, you will inevitably find that you have a following.

DO THE
RIGHT THING

Cadet Honor Code:

A cadet shall not lie, cheat, or steal, or tolerate those who do.

This is a tenet ingrained from the first moment I stepped off the bus and walked through the walls of gray at West Point.

Early on, cadets are introduced to the Cadet Honor Code. The code requires integrity in word and deed. A cadet found to be in violation of the code is brought before a cadet-run Honor Board. If merit is found, the violator's case would be reviewed by the superintendent, and in some cases, the cadet would be separated from the Academy.

To aid our moral development, honor classes were held, often taught by upperclassmen and sometimes by our tactical officers (commissioned officers assigned to the Department of Tactics). An ethically nebulous scenario was described, and we would be tasked with resolving the issue by applying honorable principles.

When the academic year began, we were expected to not cheat during exams. At the end of an exam, the professor would bellow, "Cease work!" At that moment, we were to immediately put down our pencils and comply. We also were taught to acknowledge any additional help we received on a paper or project. Acknowledgment statements would be added to the front of a paper citing the cadet who may have helped or assisted. The acknowledgment statement was so ingrained in a fellow West Point friend of mine that in graduate school she submitted an acknowledgment statement on her first paper. Her professor was perplexed but also impressed with her personal integrity.

Having an Honor Code allowed cadets a good deal of freedom. We did not have locks on the doors of our rooms. We did not need to worry about anything being stolen from our rooms. We learned to trust one another, which was foundational to building camaraderie and esprit de corps.

Our training as Fourth Classmen demanded personal responsibility. Initially, we were held responsible for our uniforms, our rooms and our weapons. And as I have previously detailed, the Academy imposed Fourth Class duties such as delivering upperclassmen's mail and laundry. We also called minutes, standing at attention and shouting for our leaders' benefit, the proper military dress for the upcoming parade or an event such as dinner formation.

At the time, I often felt those duties were unnecessary ways to possibly screw up and get in trouble. At times they felt demeaning, yet they were all part of the system to teach us personal responsibility with integrity. We learned to do the work assigned no matter how trivial and to complete it wholeheartedly. We were learning conscientiousness, dependability and a work ethic. In time we learned, as The Cadet Prayer states: "To do the harder right, instead of the easier wrong." We could be counted on to do the right thing.

Somewhere between the learning and the doing, the message of *"my word is my bond,"* became a part of my being as a cadet.

The Cadet Prayer

O God, our Father, Thou Searcher of Human hearts, help us to draw near to Thee in sincerity and truth. May our religion be filled with gladness and may our worship of Thee be natural.

Strengthen and increase our admiration for honest dealing and clean thinking and suffer not our hatred of hypocrisy and pretense ever to diminish. Encourage us in our endeavor to live above the common level of life. Make us to choose the harder right instead of the easier wrong, and never to be content with a half-truth when the whole can be won.

Endow us with courage that is born of loyalty to all that is noble and worthy, that scorns to compromise with vice and injustice and knows no fear when truth and right are in jeopardy.

Guard us against flippancy and irreverence in the sacred things of life. Grant us new ties of friendship and new opportunities of service. Kindle our hearts in fellowship with those of a cheerful countenance and soften our hearts with sympathy for those who sorrow and suffer.

Help us to maintain the honor of the Corps untarnished and unsullied and to show forth in our lives the ideals of West Point in doing our duty to Thee and to our Country.

All of which we ask in the name of the Great Friend and Master of all.

Colonel Clayton E. Wheat
Chaplain, USMA 1918-1926.

Somewhere between the learning and the doing, the message of "my word is my bond," became a part of my being as a cadet.

Later, when serving in the Army, a fellow graduate, Lieutenant "Gray," was confronted with an ethical dilemma when submitting a vehicle readiness report. The operations officer for the battalion wanted him to change the report to reflect a higher readiness than reality. LT Gray was appalled that someone would suggest changing a report. He then stated that he was unwilling to change the report despite being unsure about how that decision would impact his Army career. At that moment, he decided to "do the right thing" even if it cost him.

The interesting thing was, once Gray stated clearly that he would not back down, the operations officer changed his tune and no longer pressed him to change the report. Although this episode was disheartening for this young officer, Gray was satisfied that he had not compromised his ethics.

Sometimes, despite our best efforts, some in leadership will work against what may be best in a situation such as providing senior leadership inaccurate equipment readiness reports. LT Gray, though, chose to do the right thing.

As a leader in the Army, I believed in being candid and forthcoming with my soldiers. In return, I believe that I built trust with my subordinates. If I told a soldier I was going to look into a pay, leave or promotion issue, I meant what I said and I acted upon it. However, I also learned that I could not blindly accept what was sometimes communicated to me. For example, a mechanic completed a job order and reported the repair complete, yet when the customer picked up the item and found there was still an issue, that soldier and his leadership had to explain themselves to me. The next time that mechanic claimed to have completed a job, I would verify that his leadership had signed off on the work signifying the repairs were fully complete. I had to learn to *trust but verify*. Doing the right thing then goes beyond modeling the behavior. It also implies holding others accountable. And sometimes doing the right thing relates to personal interactions that may be difficult, yet necessary.

Doing the right thing then goes beyond modeling the behavior. It also implies holding others accountable. And sometimes doing the right thing relates to personal interactions that may be difficult, yet necessary.

A few years ago, as I previously mentioned, my classmate Chip was convalescing after spinal surgery at the local VA hospital. I again met him for dinner. We had both attended our classmate LT Gray's funeral several years after graduating from West Point. My friend Chip had been Gray's roommate at the Academy. Having known Gray for some time, I thanked Chip again for being present during that difficult time. I said, "You must have been good friends." I had assumed that was the reason he had made the effort to drive such a long distance to the funeral. Instead Chip said, "No, I never liked Gray." Surprised, I inquired as to why he did not like him.

He went on to explain that LT Gray was difficult to live with and, in his words, was "persnickety and fastidious."

I smiled, having known Gray for fourteen years. I would say he was a true engineer and liked things in a certain way and would in no uncertain terms make his opinion known. Yet, I was a realist. Not everyone loved his style.

So, I asked, "If you didn't like LT Gray, then why did you come to his funeral?" My classmate was quiet and then said, "We are West Pointers, that's what we do." I instinctively knew what he meant. We were taught and continued to model how to do the right thing even when it may have cost us to do so. This principle translated into my leadership and consulting practice.

Several years into my corporate career, I was employed by a large company as an internal consultant. I enjoyed my role immensely. However, I learned that several positions in my department would be eliminated. We were told to continue working hard in our jobs while leadership decided who would stay and who would go. Of course, the uncertainty of not knowing whether or not your job would remain created angst and some questionable activities in our department. Co-workers began bad-mouthing others in the hopes their roles would be preserved. Others maneuvered behind the scenes to ensure they had a back-up job in another department. I decided to just continue to do my job as well as I knew how.

As part of my consulting role, I met with a senior vice president for a follow-up coaching session. We worked on his development for about forty-five minutes. He gained a great deal of insight and wanted to continue the work, asking to schedule our next meeting. I told him I would like to do that but was uncertain about whether we could meet again as my position was in question. Being in another department, he was unaware of the downsizing. He leaned forward in his chair and commented, "And here you are giving me all this beneficial coaching while you do not even know if you will have a job next week!" He went on to say, "I am not sure I could do that if I was in your position." In all honesty, I did not enjoy living in a form of purgatory during those weeks of uncertainty, yet as a West Pointer, I was determined to do the "harder right."

But doing the harder right, while ethically satisfying, is not always easy and may come at a cost. My daughter Gwenyth attended art school, graduating with a Bachelor of Fine Arts in photography and film-making, with a minor in painting. Soon after graduation she launched a film company. Gwenyth combined her love for people and storytelling through the medium of film. As a young entrepreneur she worked with a variety of small businesses and government agencies and she filmed weddings. Weddings can be time-consuming and stressful events, but they can also be highly profitable. Working weddings often paid for her new cameras, lenses and lighting equipment.

Gwenyth filmed an entire wedding for one client and then, as was her custom, she backed up the video on a hard drive. However, the hard drive failed and suddenly she was looking at the possibility of losing all the footage. As you can imagine, Gwenyth was in a panic as she feared telling the bride that she may have lost all those special moments of her big day!

Initially she thought about simply giving the bride and groom their money back. However, upon deeper consideration, she determined to do the harder right. She had a written contract and needed to uphold her portion of the agreement.

She thought deeply about the dilemma and then began to lay out some ideas. "I could call the manufacturer of the hard drive and see if they have a solution or I could talk to my techie friends who might know what to do." On her own, she investigated all those options. After a few calls, she found a firm that would take a look at her hard drive. They informed her there was a slim possibility of restoring its contents. However, to use this company's services would cost her all the profit for the event.

Despite these facts, she never hesitated. Gwenyth sent the hard drive off in the hopes of it being salvaged. She called the bride and explained what had happened but assured her that she was doing everything she could to rectify the problem. She also told the newlyweds that the possibility existed that the film footage was irretrievable.

After several weeks and several hundred dollars, a technical team was able to recover the footage and Gwenyth was back in business.

Despite making no profit, this young business owner was glad she did the right thing. She gained the personal satisfaction of completing her contract and gained the self-confidence to continue to make wise, ethical business decisions. (And her mom was very proud of her!)

There may be times when you feel pressured to shield the truth, doctor a report, or try to renege on a contract. In the end, we pay for those breeches. I believe it erodes our self-respect and inhibits our ability to build trusting relationships with others. Our influence is diminished and in time we may be found out. We need only look at the daily news to see the consequences for failing to do the right thing. That is why I encourage you to embrace integrity and candor. Both are needed when playing the long game of life.

Embrace integrity and candor.
Both are needed when playing the long game of life.

CULTIVATING CHARACTER

Doing the right thing begins with holding ourselves to the highest levels of personal integrity. It requires us to choose "the harder right instead of the easier wrong."

Being an effective leader requires us to build trust with others, our subordinates, our co-workers and our superiors. If we fail to be forthright, it is never long before that interpersonal trust is eroded. In fact, it takes much more effort to build trust after it has deteriorated.

"Integrity is telling myself the truth. And honesty is telling the truth to other people."

Spencer Johnson

Therefore, I encourage you to choose the better path, a path of personal integrity and professional ethics in all you do. By doing so, you are fully accepting responsibility for every facet of your life. This includes becoming responsible for your attitude and actions in all endeavors. Only when our words align with our actions do others choose to follow our example.

Once you embrace total responsibility for your life, there are truly very few ethical dilemmas that you may need to confront.

However, when facing an ethical dilemma, let me suggest the following:

- Take a long view at the situation. If you choose to violate your personal morals, imagine how you will feel about that decision tomorrow, next week, in a year or in ten years? Will you regret making that decision? If so, then choose the better alternative.

• Reflect on a time when you had violated your personal mores. What type of feelings did that decision create? How did that breech in ethics impact your self-confidence? If you are like most people, your self-concept may have wavered. Now, think through the current dilemma you are facing. Imagine that for this one decision, you decide to violate your conscience. Imagine how you will feel as a result. Next, imagine how you will need to explain your decision to those closest to you. Then ask, is it worth it? My guess is that you will conclude that it is not worth the fallout or the emotional pain.

Instead, choose to do the right thing and live with little or no regret!

PROBLEM SOLVING

Leaders are paid to solve problems, whether they be tactical or strategic, resource allocations or human resource issues. In my years of working with a variety of leaders, I am often amazed by how tangled they can become in sorting out an issue or challenge when it appears too close to the chest for them.

Early in the history of West Point, Colonel Sylvanus Thayer was named the school's superintendent in 1817. One of his first acts was to revamp the cadet curriculum to make it engineering-focused. At that time in U.S. history, officers with civil engineering skills were needed to build fortifications, roads and bridges for the U.S. military. Because of Colonel Thayer's efforts, West Point became the first engineering school established in the country. In fact, USMA graduates were responsible for building many of America's roads, canals, bridges, and railroads.[1]

Moreover, engineering teaches critical thinking and deductive reasoning. So, it comes as no surprise that I developed a rather logical means for evaluating problems. However, as I mentioned earlier, the West Point leadership experience was by its nature experiential and often included many lessons taught by fire, so to speak. However, in this case, also by water.

One of the physical education requirements Fourth Classmen were required to pass was Plebe swimming. There was a particular requirement

---

1   West Point in the Making of America. (n.d.). Retrieved June 28, 2018, from http://americanhistory.si.edu/westpoint/history_2.html.

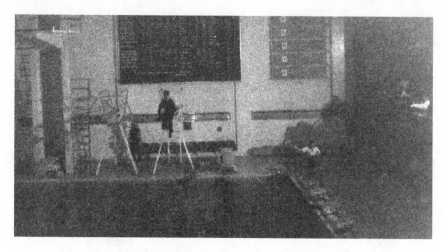

Cadet with backpack jumping in the fifty-meter pool, 1980;
Photo courtesy of Lorraine Lesieur, USMA '83

that challenged me to the core. It was called the 50-meter "Bob and Travel."

Plebes donned olive-green fatigues and oversized men's boots and hoisted a rucksack full of 40 lbs. of bricks on their backs. The task was to jump boots first into an 8-foot deep pool, land on your feet, push forward off the bottom of the pool to reach the surface, take a breath, and then make your way across the 50-meter pool, bobbing and traveling.

At the time, I was a mere 100 lbs. and so when I first jumped into the water, I experienced physics. You guessed it. I was pulled on my back and then struggled to get out of the rucksack and not drown. I barely escaped before I came up for air. (I should also add the Department of Physical Education [DPE] instructors were known for allowing Plebes to take in some water before they "assisted" and yanked us out of the water.)

My first approach to the event failed. At that point, I was not sure whether I could ever get across the insurmountable 50-meter length of the pool. So, the next time I leaned forward when I jumped off the side of the pool. This time I did not fall on my back, and I was able to get my feet underneath me and then push off the bottom of the pool. However, I was barely able to clear my nose to take in a full breath when I briefly met the surface. I knew I did not have enough air to go to the bottom again, so again, I failed. I struggled out of the dead weight of the rucksack and climbed out of the pool.

By this time, most of my male classmates had finished the event successfully and I felt defeated. However, one does not "get out" of any of these requirements, so I gave it another try. This time, I was more confident. I knew I was not going to die from lack of oxygen, so I jumped in, leaning forward, and hit the bottom. Then with both feet and all my strength, I pushed with everything I had and cleared the water to take a breath. As I descended, it occurred to me that I could make more progress by walking on the floor of the pool than by bobbing to the next descent. So, that is what I did. I walked three steps and then pushed off for the surface. I continued doing this until I reached the end of the pool.

Dripping wet as I pulled myself and the heavy sack of bricks out of the pool, the DPE instructor stated, "I have never seen anyone do it that way. Good job, Cadet!"

At the time, I remember thinking, "Well, that is the only way I could have gone that far with half my body weight in bricks weighing me down!"

When I was failing, I was actually figuring out how to solve the problem.

I relate this story not to imply I was some super woman; no, indeed! I had failed twice before I succeeded. Yet, when I was failing, I was actually figuring out how to solve the problem! I was experimenting with what I had to work with and learning how to negotiate the obstacle in real time.

I could not change the rules of the game. Yes, the men also had 40 lbs. of bricks on their backs. This was most likely only 1/5 or 1/6 of their body weight. I also could not change the fact that I had to meet this problem head on. I simply worked with the resources I had at that given moment.

It is much the same with leaders. We are forced to solve problems in real time with the resources we have at our disposal. Often a situation can seem stacked against us, but failure is simply not an option. Yet, when we make a first attempt and fail, we may not recognize that we are beginning to problem solve by learning what is *not* working first. The most skilled leaders I have worked with recognize that when approaching what appears to be an insurmountable problem, they will figure it out. They accept that they may just have to look at the problem from a different perspective. Yet, they also do not shy away from choosing a course of action, and as they get feedback, they make course corrections.

When we make a first attempt and fail we may not recognize that we are beginning to problem solve by learning what is **not** *working first.*

I coached a very talented young vice president, "Mary," an accountant in a Fortune 50 company. Identified as a high potential officer, I met with her regularly for several months. Though she was originally in the financial department, leadership was grooming her for a more strategic role. Mary was put in charge of research and development for the company. This position required her team to assess and recommend innovative energy products, as well as acquisitions of green energy companies for leadership to consider. Given the risk-averse nature of the organization, this was like asking an investor who was only comfortable with the returns from U.S. Treasury bonds to invest in Bitcoin.

I was not surprised to hear that Mary was having difficulty selling her ideas internally. She was concerned that if she did not do well, her advancement in the company would stall.

I asked Mary, "When have you ever seen someone who lacked authority influence the person who had the ultimate decision-making power and do so successfully?"

She thought for a moment (always a good sign for a coach) and then smiled and said something like, "It's like when my mom wanted to redecorate a part of our house when I was growing up and she knew my dad would say no. So, she would go ahead and get all the cost estimates and determine what the increase in value of the home would be with this improvement and then present it to my dad. Then he would inevitably agree to the renovation."

So, I asked, "How can you apply what you learned from your mother in that situation to the challenge you are currently facing?"

As a coach, we also begin to see when the wheels in our client's brain start turning. Mary then began to rattle off how she could build a business case for several of the sustainable energy alternatives she believed had merit.

Learning how to **think** about *a problem from another perspective is where we need to begin.*

Often, as leaders, we get *caught up* in a problem, yet learning how to *think about* a problem from another perspective is where we need to begin.

Checking in with past experiences for examples that we have seen others use can unlock possible solutions. In fact, those ideas can even come from completely different experiences in our lives, such as Mary's memory of her mother's approach with her father.

There is also increased risk when leaders choose to ignore or put off solving a problem.

Shortly after the great recession, I was working with a large financial institution. The leadership of the company understood that eventually a major stream of income would dry up, and this highly regulated institution would need to establish new revenue generation.

I remember asking one of the SVPs, "What keeps you up late at night?" And she replied, "Losing the refinance market and having nothing in the works to replace it." Despite this accurate assessment by this leader, the senior staff delayed planning on this issue. Instead, they hunkered down and became siloed, focusing on their individual department challenges. Not surprisingly, within one year that leadership team was forced to begin a series of layoffs in several departments to meet their revenue and shareholder targets. They avoided dealing with an issue and it cost the company a great deal.

If these leaders had faced the problem as a leadership team and experimented with new offerings, they might have identified several options to choose from and develop. Instead, a decision was forced upon them, and downsizing severely affected the morale of several departments. As mentioned previously, in the chapter on Camaraderie, had the leadership team joined forces to solve the revenue issue and held one another accountable to finding a solution, they might have shared scarce resources to maximize efforts to solve the issue. And the end of the story might have been completely different.

Several years ago, I had been developing and teaching a series of leadership programs. In that process I purchased and read numerous leadership books. One book even arrived with a coffee cup that on one side read:

No problem is unique

And, on the opposite side of the mug, it read:

Just work the problem

I stored the cup in one of my kitchen cabinets and did not think much more about it. Many years passed and I found myself downsized from a major role in a company. To make matters worse, I had recently been in a car accident that left me with a broken collar bone. I was in physical pain, feeling despondent and demoralized. One morning, I reached for that mug, read the inscription and realized that this was all I had to do.

I accepted that unemployment was a problem, but it had happened to many people. My clavicle would heal, but I could still type with one hand and make more phone calls. I had transitioned between jobs before, and I knew what I needed to do to find my next role. So, I just needed to work the problem. As I moved to action, I accepted that the situation was not permanent and that, in time, I would figure it out. And I did! My next role was more lucrative and enjoyable than the last.

I learned problem solving early—even as I was gasping for air in the Academy's fifty-meter pool.

CULTIVATING CHARACTER

When we are facing an immobilizing problem, we must first begin by believing that the issue is not unique. Rather, there exists a wide variety of possible solutions to the problem. After accepting this as fact, hope can emerge. Our job, then, is to figure out a solution from a myriad of possible options.

With this premise in mind, consider a major problem or challenge you are currently facing. Now, consider the following questions. These inquiries are designed to reframe your thinking, so you can consider several possible solutions.

- What seems to be the main obstacle? What concerns you most about …?

- How have you handled something like this before? What was the outcome?

- How have you seen someone you respect handle a similar situation?

- What part of the situation have you *not* explored?

- What other options can you now consider?

Then, based on your responses, select your preferred option and develop an implementation plan that identifies the resources you need, the time-frame to implement the solution, and the expected outcomes from your selected course of action. It is also important to break down your plan into manageable action steps so as not to become overwhelmed when tackling a major challenge. Last, build in some form of accountability for the implementation through a coach or a supportive colleague.

Particularly for more complex issues, building in an accountability partner helps to keep you focused through completion of the plan.

Leaders must lead the way and not avoid addressing issues. A leader can also modify this method for use with their team. Asking the questions above teaches them to own both the problem and solution. Do not procrastinate when issues arise, the cost of not attempting to solve a problem is always greater than any attempt.

Using this approach with several clients, and even when working on my own challenges, has resulted in some very innovative solutions as well as significant and meaningful progress toward stated goals.

NO WHINING

Cadets are notorious complainers. When we are Plebes, we complained about our outrageous squad leaders and those overbearing upperclassmen. During the academic year, we complained about those overly demanding professors or those horrendous assignments like the infamous "Sosh" paper.

If we were more lovers and poets than those more gifted in math and engineering, we made known our disdain for classes like Differential Equations and "Juice" (electrical engineering).

If we espoused a love for regiment and the military profession and did not get a slot as a Cadet Basic Training Squad leader, we complained about our assignment at the "lesser" Camp Buckner. We

The "Sosh" Paper

The USMA Social Science department conjured up a requirement for a fifteen page research paper for our International Relations course. The paper accounted for a major component of the class grade.

In addition to being notorious complainers, cadets can be incredible procrastinators. And many of us, because of our busy schedules, put off starting the paper. Many do not finish the paper until the day it is due and then make a mad dash to turn it in by the deadline.

That dash is currently referred to as the "Sosh run." Cadets don costumes as they deliver their paper. To view this phenomenon in action, visit: https://vimeo.com/124820659

complained about being forced to stand at attention for long periods of time during an inspection or before a parade.

Most of us cadets complained that we never got enough sleep. And most of the Corps would agree that the DPE Instructors had it in for all of us.

If you were a woman at West Point, you complained about the male cadets who seemed to "have it in" for you or who belittled and ridiculed your every move. We complained about those who made female cadet jokes, such as that all women members of the Corps have "Hudson Hips" disease – a reference to the wide Hudson River upon which the Academy is located. If you were male, you may have complained about women cadets dropping out of runs, being granted favoritism or degrading the standards of the Academy.

We all complained, yet I find this is a common human condition. Much of the experience of the Academy, however, was to learn to cope despite being stressed, frustrated, marginalized, overly tired, cold and sometimes even hungry.

In admitting we had no excuse, we were learning to accept personal responsibility.

The incorporation of "no whining" began early. As New Cadets, when corrected on our uniform or lacking Plebe "poop" or knowledge, we simply learned to state: "No excuse, sir (or ma'am)." In admitting we had no excuse, we were learning to accept personal responsibility.

The demands placed on cadets, whether physical, academic or emotional, were designed to develop our character and commitment to service to our nation. Those trials developed our ability to handle multiple responsibilities at the same time and to do so with candor and integrity. Although we moaned and derided the bane of our very existence, we gradually learned to handle more and complain less. We were learning to get busy and get going.

At the very end of our seven weeks of Cadet Basic Training or "Beast," our cadre would prepare us for a week at Lake Frederick. Our squad leader would build this "dream" of Lake Frederick as the land of milk and honey. A place where we could get "boodle" (access to candy and cookies). We began the week with a twelve mile road march out to the lake. We knew this week would be challenging and yet, once completed, we would then march the twelve miles back to the Academy and no longer be New Cadets but accepted as members of the Corps.

We had been warned that part of this event was to successfully complete the bayonet assault course. All summer we had drilled for hours with our M16s with bayonets attached. We would run in formation to the drilling field chanting words like, "Blood, blood, blood makes the grass grow."

In the sweltering heat of the summer, we were taught to "parry" and "thrust" to administer "butt strokes" and "slash" maneuvers. We were paired with other cadets and practiced slashing and slicing with a pugil stick (a padded stick with large oval cushions on either end). We were taught to jab at an opponent, and they in turn could block us and even overcome us.

I remember that my fellow woman opponent and I struggled to go on the offensive. We each would turn our backs and absorb blows by one another rather than attack. Aggression was contrary to how we were raised. We tried not to laugh after being yelled at as we tried again and again to fight aggressively.

We were not only being taught hand-to-hand combat, but also a warrior-type spirit. "Blood, blood, blood makes that grass grow!"

I struggled emotionally with repeating the harsh cadence as we marched to the bayonet assault practice field. Yet, after many hours of drilling, we were instructed to simulate killing our arch rival's mascot, the Navy goat. It was then, I confess, that the transformation happened. I became Athena the warrior! I screamed and yelled, and I enjoyed slashing and smashing that imaginary farm animal.

At Lake Frederick we were required to successfully navigate the bayonet assault course while running and toting the pugil stick. The course began by running up a steep mountain pass, then navigating through uneven terrain and assaulting a series of adversaries. The "enemy" in this case was rumored to all be members of the Army football team.

When you are 5'2" and weigh 100 lbs., you get worried that you may not do well fighting a football player. To make matters worse, Lake Frederick did not turn out to be the promised land. It was cold and rainy. And although my roommate, Tracey Brown Curley, and I carved a ditch around our tent, the rain still ran into our tiny living quarters. Cold and wet, we did not sleep while cradling our weapons on the night before the event.

As we prepared for our assault, I was exhausted, cold and fearful of what lay ahead of me.

I looked up at the steeply graded hill. Gullies of muddy water ran down the side of the incline. I knew this would make the climb that much more difficult. I found myself literally shaking in my boots.

Suddenly, quite extemporaneously, a classmate got up on a rock and started reciting General George S. Patton's famous speech portrayed in the movie *Patton*. In the movie, George C. Scott, playing Patton, stands in front of an immensely huge American flag. He delivers the speech to the Third Army on the eve of the Allied invasion of France:

"Men, all this stuff you've heard about America not wanting to fight, wanting to stay out of the war, is a lot of horse dung. Americans, traditionally, love to fight. All real Americans love the sting of battle.

"When you were kids, you all admired the champion marble shooter, the fastest runner, the big-league ball players, the toughest boxers. Americans love a winner and will not tolerate a loser. Americans play to win all the time. Now, I wouldn't give a hoot in hell for a man who lost and laughed. That's why Americans have never lost and will never lose a war. Because the very thought of losing is hateful to Americans..."[1]

I had heard the speech before as my Dad, a World War II Veteran, would watch the classic film on occasion. Given the reality of my current predicament, the rallying cry from my classmate began to lift my spirits. Indeed, the entire company of New Cadets began to shake their heads as he recited those infamous words. And I began to think, I can do this, I am tough, I am a warrior, I am a conqueror!

When it was my turn, I ran up the steep and jagged mountain incline. I slipped on the muddy, rocky terrain but made it to my first burly opponent of the course. I attacked, hitting my adversary's pugil stick. The cadet quickly deflected my slash with such force that I was thrown onto my back and landed in a muddy pool. He screamed at me to get up. I got up, shaken but determined. He shouted at me to do another maneuver. Even with all my effort, I again was thrown on my back in a pool of goo.

It was then, lying on my back, that I got mad — really angry. This felt a lot like my older brother, Dave, picking on me. Inside I was screaming, "You ass, that's enough!" I jumped up and attacked with everything I had within me, only to be pushed aside. But this time I was not thrown on

---

1   American Rhetoric Movie Speeches. (n.d.) Patton 1970. Retrieved March 26, 2018, from http://www.americanrhetoric.com/MovieSpeeches/ moviespeechpatton3rdarmyaddress.html

Cadets awaiting their turn at the Bayonet Assault Course, Summer 1979.
Photo courtesy of Jan Tiede Swicord, USMA '83

my back. The cadet let me go on to my next obstacle and fight my next adversary. I never did very well with subsequent encounters. However, I also never stopped. Then, before I realized it, the "test" was over. I had no broken bones. I was intact. I had completed the event and could stop and rest. I had overcome!

Some of this "get on with it" attitude came from my dad. I remember earning the right to call home while in Beast Barracks. I wanted to vent, to detail the abuses and be given comfort and maybe to be told, "It's OK, Sara, you can quit. You can come home." But I never heard that from my dad. He would listen for a bit and then inevitably say, "Well, when the going gets tough, the tough get going." And then if I kept up my tirade, he would interrupt me and simply state, "Well, this is getting to be an expensive phone call." I knew then that the call would be cut short.

To some, my dad may have sounded a bit callous. Yet, he was a quintessential member of the "Greatest Generation." He was

L. Paul Fotsch, the author's father

a Depression-era baby, the eldest son of eight siblings. Two sisters died, one as a child and the other at just twenty years of age. As children, my father, a younger brother and sister contracted diphtheria. His sister died, and my dad's growth was stunted. He was a small man standing at just 5' 4." The demands of a rugged, manual farm life made him strong and nimble. He recalled being picked on in the Army but easily flipping his assailant over his shoulder and pinning him to the ground. He would retell these stories and let us know that his tormentors never chose to fight him again.

My dad served as a medic in Patton's 7th Army, 3rd Infantry Division. He served in North Africa, Italy, France and Germany. He earned a Silver Star, a Bronze Star and four Purple Hearts, evidence of both his valor and ability to recover multiple times from injury to continue the fight. I remember seeing a long, ugly scar running along the outside of his right thigh, evidence of where pieces of enemy shrapnel had hit him. And like many of his generation, he never spoke much about his war experience or his valor. It was not until much later that I understood more about my Dad's service. At one point, he had been separated from his unit and was missing in action (MIA) for a couple of weeks before he found his way back to a friendly unit. He had his shortcomings, like we all do; nonetheless, he had daunting drive. He earned three bachelor's degrees, four master's degrees and a PhD. The eternal student, he was always reading several books at one time. His admonition to hunker down and keep going, was something I did not want to hear at the time. But, his words were exactly what I needed to hear.

His admonition to hunker down and keep going,
was something I did not want to hear at the time.
But, his words were exactly what I needed to hear.

Later in life I met another hero. Through the West Point Society of Richmond, I had the privilege to get to know retired U.S. Navy Commander Paul Galanti. Paul is a Naval Academy graduate and former Viet Nam prisoner of war (POW). Paul became a real champion to me; he encouraged me and mentored me in my various corporate roles.

At just twenty-six years of age, Paul, a Navy pilot, was shot down by enemy fire and spent nearly seven years as a prisoner. He was held in the infamous prison camp known as the Hanoi Hilton. He was tortured and beaten; even so, he endured until his release. As I got to know him, I inquired how he kept going given the unrelenting persecution of his captors. He thought for a moment and said, "Well, I knew that the CAG (Commander Air Group) James Stockdale was in the camp. He was forty-six years old and was badly injured with a shattered knee." Paul, continued, "I thought, if this 'old guy' can do it, then so can I."

The endurance of his leader was enough for Paul to make it through another minute, another hour, and then weeks and years. In turn, Paul and the other detainees found ways to communicate with one another to keep up their spirits despite their situation.

The Tap Code

TAP CODE	1	2	3	4	5
1	A	B	C/K	D	E
2	F	G	H	I	J
3	L	M	N	O	P
4	Q	R	S	T	U
5	V	W	X	Y	Z

- Letters were comprised of two sets of taps, separated by a pause. The first set goes down the column. The second set goes across the rows.

- Example: For "A," tap once, pause, then tap once more. For "B," tap once, pause, then tap twice. For "S," tap four times, pause, then tap three times.

- "C" and "K" shared a block because of their similar hard sounds.[2]

First, using the Tap Code method, inmates could pass information and encouragement to their fellow captives. The code was simple and easy to learn. The prisoners would tap on walls, pipes and even one another. They used abbreviations such a "GN" for good night and "GBU" for God bless you. [3]

---

2   Metal Gear Wiki. (n.d.) Tap Code. Retrieved March 26, 2018, from http://metalgear.wikia.com/wiki/Tap_code.

3   Ibid.

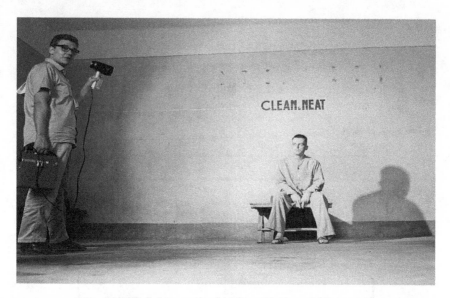

POW Paul Galanti was forced by his captors to take this picture.
Photo used with permission of Paul Galanti

Later in their internment, they used a form of sign language to more quickly communicate undetected. Using both communication systems enabled the inmates to bond and maintain their sanity. It also created a sense of order by re-establishing the chain of command.

As the years passed, Paul and his fellow prisoners were eventually allowed to spend time in larger groups. To fight off boredom, they began to teach one another difficult subjects such as chemistry or would re-create movies in great detail and all by memory. This type of learning gave Paul an insatiable desire to learn when he was released. Moreover, Paul came through his captivity with the ability to encourage and inspire others. One of his quotes puts most of our daily challenges in perspective:

"There is no such thing as a bad day when you have a doorknob on the inside of the door."

Paul Galanti, U.S. Navy Commander (Retired)

Another benefit of knowing Paul was meeting his equally impressive wife, Phyllis. While Paul was fighting his own battles as a POW, Phyllis would emerge as the chairwoman of the National League of Families of American POWs and MIAs in Southeast Asia.

Phyllis had gone to college to become a French teacher; however, she was so shy, she was too intimidated to teach. After Paul was taken captive, she began a tireless effort to free the POWs. She was the first non-elected official to ever address a joint session of the Virginia General Assembly. She led a powerful letter writing campaign to raise awareness across the country concerning those husbands and fathers still in captivity. Remember this was in the late 1960's; raising awareness in this manner had never been done. Nevertheless, the pioneering women of the National League of Families eventually earned the support of the nation. The letter writing campaign even led to a meeting with President Nixon and Henry Kissinger.

The business tycoon, Ross Perot sponsored trips of delegations of these wives and family members to the Paris peace talks. On one trip, Phyllis learned of a secret meeting of North Vietnamese. She bolted into the conference room and in perfect French, including some expletives, demanded the North Vietnamese improve the conditions of the captives and release the American POWs.

Her tireless efforts strengthened the hearts of the women and families left waiting and led to the release of many POWs, including her beloved Paul. Her transformation, as described by her husband, was summarized like this, "When I left, my wife was a kitten. When I came back, she was a tiger!"

"When I left, my wife was a kitten.
When I came back, she was a tiger!"

Paul Galanti, U.S. Navy Commander (Retired)

We lost Phyllis a few years ago to an aggressive form of cancer. At her funeral, a good friend delivered her eulogy that detailed her notable

Paul and Phyllis shortly after his release.
Photo credit: Richmond Times-Dispatch, February 15, 1973

achievements both before and after Paul's release. One memorable tidbit shared was that Phyllis's chosen email address was "nowhining@ comcast.com."

When I heard that, I smiled, but I was not in the least surprised. Phyllis, Paul and my dad demonstrated fortitude over complaining. They knew there was no upside in lamenting their situation. Rather, they continued the fight and chose to embrace hope despite the reality of their situation.

Just as I learned as I was struggling to my feet from that pool of mud, I chose to get up, keep one foot in front of the other and just keep going … no excuses.

CULTIVATING CHARACTER

If we set out to achieve anything of substance in life, most likely we will face setbacks and discouragement. However, our internal messaging is a huge factor in how we can manage these painful and stressful scenarios.

A leader can only instill optimism in followers if he or she maintains motivation. Leaders then must be expertly attuned to their inner voice while also assessing their team members' attitudes and behaviors.

We must begin by becoming acutely aware of our self-talk. *Do you have a self-defeating monologue?*

Begin by quickly diagnosing when you begin to hear that internal negative voice. Note what sets off a downward spiral into negativity. Become aware of who you are with and what circumstances typically elicit destructive thinking. This information can inform you what situations are best to avoid or where to find a work-around.

An antidote to breaking self-destructive messaging is to insert in its place re-affirming language. Here are some suggested methods:

- Seek out podcasts or video recordings of speeches that transport or inspire you.

- Amp up music that gets you motivated and uplifted.

- Read biographies of people who have achieved a great deal in life. You will learn what they said to themselves when hope seemed to be lost or the odds of success were dim. Just as Paul Galanti and his fellow captives used the Tap Code to communicate and encourage one another while enduring years of imprisonment.

- Memorize quotes that get you focused forward. Choose a motto for your specific situation that succinctly states your attitude concerning that situation. When I need to get my mojo on, I recite one of the many cadences we learned when marching at West Point. We would sing as we ran up a hill, "Little hill, no sweat, you can do it, all the way!" In other words, you got this!

- Call that trusted supportive friend, mentor or coach. Listen as they encourage you and help you gain perspective on the situation and mitigate negativity.

Then, you too can choose "no whining" and instead get up and get going, one foot in front of the other.

HUMOR

Although we often complained, cadets developed a quick wit and a sense of humor to offset the repressive reality of our daily grind. There were many opportunities to be put in compromising positions as women. Further, there were ample opportunities to mess up as a Fourth Classman. As we relived those experiences with each other, we would laugh and our burdens seemed less dark. Our existence felt a lot less dire, rather rich and in some cases very entertaining. The lesson of learning to laugh and find the humor in daily life has kept me sane and not just at West Point but throughout my life.

The lesson of learning to laugh and find the humor in daily life has kept me sane and not just at West Point but throughout my life.

SAMI

Saturday Morning Inspection (or SAMI), as the name implies, was held on Saturday mornings. Upperclassmen would inspect our rooms for everything from how tight our beds were made to how exceptionally clean our room appeared. They would check our wardrobes and drawers to ensure complete compliance with regulations. Every sock was folded to create a "smile," and every item of underwear was folded as required and then arranged in the drawer in the specified manner.

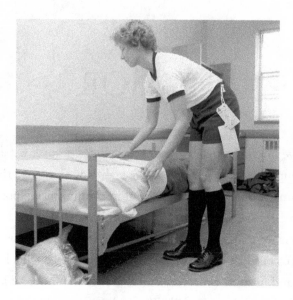

Female cadet making her bed.
Source: U.S. Military Academy Library Archives Collection

My two Beast roommates, Catherine O'Connor and Tracy Brown Curley, and I worked diligently the night before the inspection. We dusted every crevice, polished every shoe and inspected every drawer – all with the hope that our room would be found acceptable. If so, then we could make our way through the upperclassmen and their harrowing demands of recitation of Plebe poop to then stand in line to be able to make a fifteen minute call home. We had been warned that the cadre was merciless and no matter how perfect the room, they would find something wrong. Or simply because they were in the position to do so, they could tear up our beds, throw out all our carefully folded clothes and suspend all our hopes of phoning home.

As we began to arrange our room, I had a nondescript box of feminine pads and tampons. But there were no regulations describing where this item should be properly located for an inspection. Not knowing what else to do, I moved it to the back of one of the drawers and hoped the inspectors would not see it.

When the first sergeant and squad leader entered our room, they performed as we expected, tearing up our beds and finding fault. While inspecting the mirror, the sergeant sadistically smeared petroleum jelly on the mirror. Then acting disgusted, he ranted, "What is Vaseline doing here?" We knew we were powerless to challenge this incident and our hopes of getting to call home were dwindling.

Next, the sergeant began looking through our wardrobe and drawers. He then opened the drawer containing my box of "monthly supplies." Believing it was some type of contraband, he turned to me and asked, "What is this? Trying to sneak something in here?" And then, he opened the lid and his face turned red. He quickly put the box down, yelled for us to clean up the room and quickly left. Once they were gone, my roommates and I began to laugh and laugh. So, I kept the box for all future inspections. It became our secret weapon.

Even when my roommates and I believed we were at the mercy of our intimidating male cadre, they themselves were at a loss at times to know how to lead and manage women. Perhaps we were teaching them they would need to learn new skills other than intimidation and condescension. And isn't it ironic that even today, men often struggle with how to manage women in the workplace?

DINNER FORMATION

When the academic year began, Plebes were assigned to one of thirty-six companies. I and two women classmates, Pam Abear and Jenny Campbell, were assigned to H-2 (Company H, Second Regiment).

One of our Plebe duties was to call minutes prior to each formation, be it a meal or parade. Minutes began at five minutes prior, two minutes and then one minute prior to the assemblage. The assigned Plebe would stand below the clock, and at the exact time "sound off" something like the following:

> "Sir, there are five minutes before dinner formation.
> The uniform is white over gray with sabers."

This communique informed the cadets on that floor that they would don their white dress shirts and gray pants. Seniors would also add their sabers. Other uniform options could be dress gray, full dress uniform, etc.

Between the five-minute and two-minute call, the Plebe would stand at attention. He or she was then a target for upperclassmen to ask any variety of question. What's the dinner menu? Pop off the days! (That is, begin a recitation of how many days until the Army-Navy game, winter break, spring break, graduation, etc.)

My roommate, Pam, was from New Hampshire and had a thick New England accent. When she called the minutes, it sounded like this:

> "Sa, thewe awe five minutes until dinnah fowmation.
> The unifowm is white over gray with sabahs."

Immediately, several upperclassmen charged from their rooms and surrounded her. One of the cadets, pointed to his saber. Yelling, he demanded, "Abear, what is this called?" To which she would quickly reply, "Sah, it's called sabah."

This so entertained the cadre that it led to requests for her recite the minutes again and again. Back in the room, we all laughed hysterically.

Pam, being a proud New Englander, didn't allow the added "attention" to make her change her accent. She was demonstrating self-confidence, a skill she would need both at West Point and later as an Army officer.

OMAR – OH MY!

On R-Day (Reception Day), at the beginning of Cadet Basic Training, more than 1,000 of my classmates and I were marched into a large hall and measured for our numerous uniforms. The measuring was done quickly so that the tailors could get to work and the legion of new cadets could wear a uniform in the parade for our parents later in the day. We were told that some of those tailors came from New York City to help with this herculean effort. Obviously, there was no time to make fine adjustments to those uniforms. In fact, when a cadet might question a fit, he or she might be told, "Ita fita fine."

One pair of my dress pants never seemed to fit quite correctly. Once, later in the academic year, I noticed those pants came back from the cleaners and the zipper was a bit bent. We had a parade that day, and as these were the only pants I had that were clean, I quickly put them on and added all my additional parade attire.

This parade was honoring Omar Bradley, the famed World War II general. He was dubbed by journalist Ernie Pyle "the GI General" for his unassuming manner and frequent donning of a common soldier's uniform in combat.[1] General Bradley, a USMA Class of 1915 alum, eventually was named the chairman of the Joint Chiefs of Staff in 1948 and, with that, earned his fifth star. In 1980, however, he was eighty-seven years old and confined to a wheelchair.

As is the protocol for honoring dignitaries, like General Bradley, the Corps would conduct a "Pass in Review" while in full dress uniform.

---

1   Hickman, K. (2008, November 14). The GI General: General Omar Bradley. Retrieved March 17, 2018, from https://www.thoughtco.com/world-war-ii-general-omar-bradley-2360152.

Cadets at the U.S. Military Academy marching with "Eyes Right" during the Pass in Review prior to an NCAA football game; September 29, 2012, in West Point, NY. Photo credit: DOD Photo / Alamy Stock Photo

Beginning with First Regiment through Fourth Regiment, each of the Corps' thirty-six companies would execute "Eyes Right" as we passed the individual. As we marched near the dignitary, the company commander would shout, "Eyes," and immediately his platoon leaders would echo the command. Next, the commander would yell, "Right!" and the entire unit would simultaneously snap their heads to the right toward the honoree. Then after marching past the dignitaries, the commander would then command, "Ready!" and the platoon leaders would echo, "Ready." The commander would bellow, "Front!" and we would snap our heads forward in unison.

All parades are stressful for Fourth Classmen. It starts from the time before the parade begins when indiscretions found in our uniforms would give opportunity for hazing and questioning. Then, once we began to march, we were barraged with correction. "Dress right, Beanhead!" This means that you were not staying in line with the person to your right. Hence, we were under constant fear that we would screw up and be singled out. If not during the parade, we could even be disparaged for our failures when we were under the "Sally Ports" away from the view of the parade attendees.

As my company approached Omar, and the company commander shouted "Eyes Right," and I snapped my head. I felt a sudden gush of air entering my pants and realized that my zipper had unzipped. I was mortified! Fearing the upperclassmen around me, I could not (in the middle of a parade) stop to zip my pants without bringing down the forces of hell upon me.

I marched on with the air ripping through my loins until we were back in our assembly area of the barracks. Before we could be dismissed, our squad leader walked down the squad and checked us out, giving us feedback on our marching or just holding us there as long as he chose. When he came to me, he looked down. Noting the gaping hole and holding back a laugh, he said, "Cadet Fotsch, do you have something to say?" Beyond embarrassed, I asked to make a statement and said, "Sir, may I make an adjustment?" He nodded, and I whipped my hand out and zipped up those damn pants!

In not trying to zip my pants during the parade, I potentially had avoided making the company look bad. While reaching for my zipper, I might have moved out of line with those to my right or, worse, I may have tripped and my rifle might have slipped off my shoulder. Although my squad leader did not make note of my restraint, I was in fact demonstrating esprit de corps! Even when making mistakes, we can still demonstrate leadership.

WRONG ROOM

However, I seemed destined to mess up with that squad leader. And this incident probably sealed his opinion of me. As I mentioned before,

Plebes were required to "ping" and square corners even when entering a stairwell and ascending a building. We would hug the far wall as we pinged up the steps. When we reached a landing, we would square the landing before proceeding up to the next flight. This process allowed upperclassmen to have the preferred inner pathway. Despite this, and always being pressed for time, Fourth Classmen learned to mount those stairs quickly.

One afternoon I was racing up those stairs, making good time. I emerged on our floor and counted off the rooms until mine and opened the door, closing it behind me. There, sitting at my desk, was my squad leader. I was confused. What was he doing in my room? Was this a new sinister trick to move our rooms without our knowledge? The upperclassman seemed equally surprised. My mind was racing. I had closed the door and he was a male; I could get in trouble for this. The cadet then asked, "Cadet Fotsch, do you have something to say?"

In that instant I realized I had miscounted the floors and gone one floor too many. I was in *his* room! I stammered, "Sir, I think I made a mistake!" And as I said this, I did an "about face" and pinged the heck out of that room and down the stairs to my room.

Later, at dinner formation, likely convinced of my perpetual ineptitude, the squad leader stated for all to hear, "Cadet Fotsch made a mistake today." To his credit, he did not elaborate on my error to the entire platoon. In fact, he smiled and continued peppering my other squad members with Plebe "poop" questions.

Perhaps because I owned my error, he granted me a reprieve from further humiliation. How refreshing it is when we have the internal fortitude to freely state, "I messed up" or "I was responsible for that" or "I apologize." And good leaders seem able to understand that no one is perfect and do not expect that of their subordinates.

RAISINS

I clearly remember a Cow (or Junior) in my Plebe company. He was our platoon sergeant and also played on the Army rugby team. He would make sure we knew our Plebe recitations and correct us when our uniforms were out of line; however, he also made it a point to encourage us and at times even joke with us. January and February in our "rock bound highland home," as we sometimes referred to West Point, could be brutally cold and bleak. As Plebes we would lament, "The skies are gray, the buildings are gray, and we read the *New York Times* and it has no funnies!"

One extremely cold and windy day with snow on the ground, the uniform was changed from our gray coats and scarfs to our black parkas and beanies. We would pull the black, gray and gold beanies on our heads but then pull the thicker wool parka over our sweaters. Then, to properly wear this uniform, we pulled the hood of the parka over our heads. The hood then formed a kind of Hershey kiss around our beanie-laden cranium.

As five of his Plebes stood at attention in formation waiting for breakfast formation to begin, Cadet "Good Guy" walked up and down in front of us. He began to laugh as he looked at us. He smiled and asked, "What do we look like right now?" We had no idea what he was talking about. He gave us the option to guess, and of course, we were dumbstruck. As Beanheads, we were rarely asked our opinion on anything. Then he said, "Raisins! We all look like a bunch of raisins!" My classmates and I began to chuckle and smile and the severe wind and cold did not seem that bad at all. And we felt a sudden kinship with our leader, because he had admitted that he also resembled a dehydrated grape.

Just as that chuckle created a mood shift in those cold cadets, a leader can shift the mood of his team with light hearted humor. Leaders who effectively use humor can increase the trust within a team, leading to greater team cohesion and productivity.

PAY BACK

Many of you may have a younger sibling like I do. My brother Lee was eighteen months my junior and sometimes made it his stated purpose in life to give his sister a hard time and inform on me to our parents. Soon after graduation from West Point as a 2nd lieutenant, I attended Airborne school in Fort Benning, Georgia, to learn how to jump out of a perfectly good airplane. A second lieutenant is the most junior of Army officers. This rank was displayed on my uniforms with two gold bars on each shoulder board. In Army parlance, those bars are often referred to as "butter bars."

At this time my brother Lee was a junior in college and had decided to pay for college by joining the National Guard. While I was learning how to jump out of airplanes, Lee was attending Infantry Basic Training as an enlisted soldier. I decided I would be a good big sister and check up on him and bring him a bag full of cookies and candies.

Up to that point in his training, Lee had managed to stay below the radar with his drill sergeants. They did not know his name. However, when I arrived, he came out and saluted me. This action alone, as you can imagine, was very enjoyable for his sister. Lee's head was shaved and he had lost weight, so I asked him how he was being treated, to which he said something like, "Ma'am, they are taking good care of me." Then I handed him the bag of sweets, and he saluted me again and left.

Shortly after my departure, several drill sergeants surrounded my brother and said, "We don't care how much brass your sister has, your ass is mine!" And from that time forward, Lee became the object of a great deal of "extra attention" from his leaders. Apparently, he did several hundred push-ups as a result of my visit. And now at most family reunions, Lee recounts and further embellishes this story. After which, I simply say, "Payback is a sister with butter bars!"

Although my intention was to encourage my brother, I did get to assess if he was being treated well and I would not have hesitated to use my "brass" to intervene had I noted he was hurt or underfed, etc. In the end, he knew I cared about him and came to visit him, just like a leader would do.

I did emerge from the shackles of my own doing, graduated from USMA and had a successful Army career before starting my own company. However, my fiascos with the Academy continued.

LEAKING LADY

After beginning my civilian career, I opted to become a military academy liaison officer (MALO). A MALO assists high school students in their local state through the two-year process of gaining acceptance to our alma mater. The current MALO wanted to add a woman graduate to assist in recruiting women to West Point.

Becoming a MALO required a couple days of training at USMA. However, the training would take place about ten weeks after I gave birth to my second daughter, Gwenyth.

With each of my daughters, I tried to begin their life by breastfeeding which meant that daughter number two would need to accompany me to West Point. While the other MALO trainees would stay in the cadet barracks, I had to secure a place for myself and my daughter to stay.

My former cadet sponsor, Major Palmer Bailey, was still teaching at the Academy. He and his family would be away when the training took place, and they graciously offered their officer quarters during my stay. His wife also connected me with a sitter to watch my baby as I attended the instruction. During the training, I was happy to reunite with several women graduates who were also choosing to become MALOs in their states.

As I entered the gates of the Academy, some seven years after graduating, I had a strange foreboding that I could make a mistake and get in trouble, once again, at West Point. We were asked to attend wearing our officer dress green uniforms. As it was summer, I wore the short sleeve blouse and pants.

After finishing the morning session, two other women grads and I opted to go to lunch at the Officer's Club. As we were eating, there were several male officers at tables near us. All was going well when my classmate, Stacy, tapped me on the hand and said, "You are leaking!" Looking down, my shirt was wet on the left side. My breast milk had come in and run through the pads in my bra. Mortified, I thought, "Only I would have this happen at West Point."

I took my black military issued purse and held it over my blouse. We rushed to my classmate's barracks room. Always the good soldier, she was well prepared and had a spare military-issued green blouse that I borrowed for the remainder of the day. In an ironic juxtaposition, my fate as a woman with all the accompanying appendages had again gotten me in trouble at West Point. Nevertheless, we women still had that camaraderie ingrained within us!

As Mark Twain is credited with saying, "Humor is tragedy plus time." At the time, the "Omar, Oh My!" and the "Wrong Room," incidents not to mention the other embarrassing events recounted here, were devastating to my self-concept then. Yet, in time, they became fodder for laughter for me and my classmates. And who says you can't laugh over spilled milk?

"Humor is tragedy plus time."

Mark Twain

Embarrassment is inevitable. Laughing, however, is a choice. Humor in all its forms is good for the heart.

CULTIVATING CHARACTER

Perhaps you are currently in the midst of a situation that you do not find humorous or remotely enjoyable. I have found myself in those situations as well. We are told, "Happiness is a choice!" Yet, that very emotion can seem so elusive. What I began to discover at West Point was that learning to laugh is the antidote for the other option — resignation to the issue or, worse, becoming depressed and despondent.

In my years as a coach and consultant, I have studied and taught others to develop their emotional intelligence (EI). Research shows that a person's EI is a greater predictor of success than a person's IQ. And unlike IQ, which does not change much as an adult, EI can be developed through intentional practice.

One of the fundamentals of improving EI is to learn to become intentionally aware of our emotions throughout each day. In thinking about our emotions, we learn to master them rather than have them master us. I emphasize to my clients to begin by first naming their emotions as they occur. Be specific. Am I mad or just frustrated? Or am I discouraged? Am I happy or am I delighted? Or am I exhilarated? Then I ask them to record the physical sensations of those emotions and how long the feelings last. I also instruct them to become aware of which situations typically elicit more pleasant emotions and which create more negative emotions.

In time and with practice, my clients can recall, in detail, their positive emotions and those situations that elicited these feelings. They can, for example, choose to vividly recall a humorous, restful time to combat a negative situation. There is incredible power simply in learning to recall humorous events and then relive them. In managing our emotions, we can also diffuse a negative scenario and view it as transient and less defining.

There is incredible power simply in learning to recall humorous events and then relive them. In managing our emotions, we can also diffuse a negative scenario and view it as transient and less defining.

Leaders who are emotionally aware will sense when a team needs some levity, as my platoon sergeant demonstrated that his Plebes needed to laugh on that frigid morning. As a leader, you can lift a teammate with a sincere compliment or smile. You can impact a team's mood by learning to make light of a situation or, better yet, teaching by example how not to take oneself or a situation too seriously.

Here are some suggestions to awaken your humor:

1. Notify your face – Smile. Just smiling can release endorphins into your system and alleviates stress.

2. Record in detail the most humorous episodes from your life that you now view as hilarious. Keep track of how your body responds to each memory. Did you chuckle, smile and relax?

3. Now, while in a more relaxed mood, consider the issue or challenge you are currently facing. Can you view this issue with greater perspective? Can you view it as fleeting and less permanent? You may even be able to think about possible solutions when you are in this state.

4. Imagine how you might view this temporary situation as humorous in a year, two years or ten years from now. This last suggestion takes

practice, but in time, you can leverage humor to find not only levity but clarity and actionable solutions.

Embracing humor is the antidote we all can choose as we face embarrassing or trying times. It is healthy to laugh and reminisce about the ridiculous situations we have all lived. As for me, when I say, "Omar, Oh My," I always grin.

FAIL FAST

Early in Beast Barracks, within the first week, we were marched into an auditorium for a briefing. As we sat down, the officer at the front of the room began by saying, "Look at the person seated on your right, and now look at the person on your left. Two of you will not be here when you graduate."

I remember thinking while hearing this sobering proclamation, "I hope I have what it takes to get through this place."

There are so many ways that a cadet can experience failure. I certainly had that experience. When I attended the Academy, every class assignment would be graded and posted for everyone to see. After every test or exam, our grades and names would be posted. Our grades on physical fitness exams and classes were also made public. Worse for me was the public posting of our weight!

As a runner, I wanted to be rail thin. Yet, our high caloric diet of three meals a day combined with the stress of our daily life resulted in a fifteen pound weight gain even as my weekly mileage increased while training for a marathon. The perfectionist in me found this incredibly frustrating.

One could also fail competing on one of the many athletic teams at the Academy. I developed a special respect for Army football over the years. From 2002 to 2015, Army lost to Navy for fourteen years in a row! This meant that there were Army football teams that, for their entire four years at West Point, were never victorious over our arch rival. In 2015, I attended the annual contest with Navy at Philadelphia's Lincoln

Financial Field stadium. Army took an early, impressive lead and the score was very close until the last quarter. Then Navy took the lead and won 21-17. At the end of the game, as we sang the alma mater, I remember the tears streaming down the cheeks of the Army football team members. It hurts to lose! It's painful to fail.

For me, like many cadets, the greatest struggles of my years at West Point were academic. Although a straight-A student in high school, I did not do nearly as well at West Point. There were extenuating situations, at times, when I felt a professor was unfair. But other cadets seemed to perform well under some of the instructors when I did not.

In fact, I rejected the thought of pursuing an advanced degree for several years after graduation simply because I did not want to feel humiliated in a classroom ever again. Fortunately, like my dad, I had a love for learning. Instead of pursuing an advanced degree, I read a great deal and experimented with leadership methods in all my roles.

Success is a process built upon failure.

I have also had my fair share of setbacks and disappointments in my career. However, in time, I have learned to think of these events in the terms I did when I was just a young cadet. I have embraced the notion that success is a process built upon failure. The key, though, is to learn to fail fast–that is to learn to quickly recover from an unsuccessful attempt. I discovered that this ability is a common trait among some of the most accomplished leaders in history.

The USMA Class of 1915 is called "the class stars fell on." Of the 164 graduates, fifty-nine rose to the rank of brigadier general (a one-star general) or higher, three to the rank for full general and two to the rank

of General of the Army. Included in this impressive class were the likes of
Omar Bradley and Dwight D. Eisenhower, who became the 34th president of the United States.

However, there was great disparity in the academic standing of these
generals. In reading further about the Class of Stars, I was surprised
to learn that President Eisenhower was only an average student.
Academically, he graduated 61st in a class of 164.

One of the "old graduates" that I came to admire was General George
S. Patton, Class of 1909. Patton is often considered the most successful
combat general in World War II. However, Cadet Patton was forced to
repeat his entire Plebe year because he failed mathematics. Undeterred,
Patton hired a tutor and worked very hard and eventually graduated 46th
in a class of 103 cadets. Legend has it that his statue at the Academy
faced the Cadet Library as punishment for his lack of abilities in the
classroom. As I chose to study in the library, I recall passing the statue

General George S. Patton Jr. Monument at West Point

and saying, "Patton, there is no way I am taking five years to get through this place!"

Even with this failure early in his life, General Patton went on to achieve a great deal despite his reputation as being flamboyant and controversial. Known for not knowing when to shut up, he was demoted after a tremendous victory in Sicily for slapping two soldiers he believed were cowardly. Although he was disciplined for his actions, he then fought his way back into a combat command, leading the 3rd Army in France and Germany, eventually earning his fourth star.

The fact is that failure is a certainty, especially if you are attempting anything of real significance. If you haven't failed at something recently, I would suggest that you may have set your sights too low.

The fact is that failure is a certainty, especially if you are attempting anything of real significance. If you haven't failed at something recently, I would suggest that you may have set your sights too low.

Recently, I was listening to an interview of Scott Hamilton, the 1984 Olympic ice skating gold medalist. Scott was discussing his new book, *Finish First*. During the interview, Scott related a story about his son who plays ice hockey. The hockey team had lost a critical game and his son was angry and despondent. Scott then asked him, "What would you do differently if you could do things over?" His son began to declare how he might have practiced differently, prepared better, etc. Then Scott asked him a truly powerful question: *"What would you have learned if you had won?"*

His son thought for a minute and realized that he would not have learned much at all if he had won.

The truth is that we can learn a great deal from failure. It is a source of good data that we can use to figure out our next steps or our next attempt. The correct way to view failure is to view it as transient and not life-defining. Rather, failing is a learning process on our way to success.

Those who achieve the most in life learn to reframe failing. They look at the situation and learn from it. That is, they are able to let go of the past, keep moving forward with the knowledge they gained and keep getting better.

I recall suffering under a philosophy instructor in my sophomore (or "Yearling") year. To make matters worse, I had endured Captain "Stickler" in advanced English as a Plebe. As a Fourth Classman, I remember getting upset at earning a C on a paper. I went to his office to inquire about his reasoning for my grade. He began to berate my writing, which up to that time in my life, I had believed was very good. I had taken AP English in high school and had received an A+ and a 4 on the AP exam. I had even had some of my work published when I was 16 years old!

After hearing Stickler berate me for some time, in exasperation, I blurted out, "You know, I could have gone to Harvard." Irritated, he reached for a folder and took out some test scores. These were assessments we had taken when we were in the middle of Cadet Basic Training (also called Beast). Obviously, Beast was not the ideal time to take a standardized test. Nevertheless, he revealed my score and stated, "Cadet, you barely made it into advanced English." In other words, I was not very good and was fortunate to even be in his class.

Imagine my disappointment when I discovered that I would have this same instructor for philosophy in my Yearling year. By this time, I spent more time on my papers and even had an officer proofread my submissions. However, despite every effort, I would consistently get C's or low B's from this Captain Stickler.

A woman cadet in my class seemed to do well with "Stick," so I asked her how she got such good grades. She said, "The more perverted I write, the better the grade, so I keep writing that way."

During the course, both inductive and deductive reasoning were taught and we had to argue in our papers using both methods of logic. For those of you who may be several decades removed from using these forms of logic, I provide the following descriptions:

Deductive reasoning begins with a generally accepted theory or premise that you test by applying it to specific incidents. For example:

All dogs have fleas. Scout is a dog, therefore he has fleas.

The potential issue with using this form of logic is that the premise, "All dogs have fleas," is not necessarily true and could lead to a faulty conclusion.

Inductive reasoning, by contrast, is done by making a series of observations and then recognizing a potential pattern and forming a hypothesis. For example:

You observe that none of the schnauzer puppies in a kennel have fleas, but all of the poodle puppies have fleas. You hypothesize that poodles are more likely to have fleas than schnauzers.

Note that with inductive reasoning, you are not certain of your hypothesis and it can change with more observations and data.

However, despite my best efforts, I never achieved great grades in philosophy, and as I had other classes that I needed to focus upon, I endured and moved on. In my mind, I had failed. Moreover, I believed I had not learned much, if anything, from Captain Stickler.

Despite the pronouncement of that instructor, over the years I have written volumes on a variety of subjects. I wonder now if Stickler would be surprised? Later, I discovered that I had learned something valuable in his class. I had learned to apply both inductive and deductive logic to the task at hand!

About twenty years after enduring Captain Stick, I was hired as an internal consultant for a large corporation. Within a few weeks of hire, I was asked to write the diversity training curriculum for the company. I was to begin by meeting with the Executive Diversity Council (EDC). It was a white male-dominated EDC in a risk-averse company. I was to create a facilitated session, where the members would agree on the guiding principles from which to develop the curriculum.

As an aside, I have made an entire career from assuming jobs and assignments that are difficult and that I may not know a great deal about. However, I eventually figure out how to proceed and succeed. This was one such project. As I mentioned in the Problem Solving chapter, figuring out how to survive the Bob and Travel as a Plebe may have laid the foundation for me to take on increasingly more complex assignments in my career.

After some research, I discovered an article that described using inductive reasoning to uncover fundamental premises when attempting to create an inclusive workplace. I quickly recalled the inductive reasoning from my painful cadet experience with Stickler. I then designed a facilitated approach to elicit specific examples in the EDC's lives where they had experienced being undervalued, ostracized or minimized. From these findings, the EDC induced the principles of a healthy, inclusive workplace. One major discovery and agreement the group came to was that the senior leaders of the company needed to teach the curriculum. They bought into the belief that the senior leaders must first embrace these principles and then teach them to their subordinates. A leader-led training would be the best means to move the needle on becoming more inclusive.

After that successful session, a member of the EDC and chief counsel for the corporation approached me. He praised my approach and then asked if he might borrow my methodology and share it with his staff of twenty-plus attorneys. He felt that even his own staff lacked the ability to leverage inductive reasoning in their work.

I smiled and was happy to share my method. Then I thought about Captain Stickler and my philosophy fiasco. I realized that even as I was failing, I was learning. The curious thing is that I did not remember what I had learned until I was confronted with a scenario that required that knowledge. I had stored my learning from that course in the far reaches of my brain. And then, when the need presented itself, I was able to extract what was needed. I wonder, had it not been such a painful memory, would I have recalled those concepts so easily?

Failure is transient.

As Scott Hamilton so correctly stated, "What would I have learned had I won?"

The West Point experience ensured that most of us failed many times before we graduated. We were tested so often, on so many levels, that it created some self-doubt but perhaps some needed humility as well. We learned to accept failure and learn from it but not let it define who we could be.

Failure is transient. Choose to learn from your failure and make the needed changes. In this way, you will *Fail Fast* and move closer to your goals!

CULTIVATING CHARACTER

*"Failure should be our teacher, not our undertaker.
Failure is delay, not defeat. It is a temporary detour, not
a dead end. Failure is something we can avoid only by
saying nothing, doing nothing, and being nothing."*

Denis Waitley

We all fail. But to embrace the concept of failing fast, I suggest the following action steps. Again, I emphasize that one must choose to take action to avoid a defeatist attitude that can emerge after a loss.

1. Take time to process your feelings concerning that failure. Suppressing negative emotions can lead to other issues, so allow yourself the ability to mourn the loss.

2. Accept the failure. Remember that you are not alone. Everyone fails and anyone who has ever accomplished anything has failed many, many times. Accepting failure does not mean that we allow the situation to drag us down and define our future. Moving past a setback may be especially difficult to accept for those of us who are high achievers. Somehow, we think a setback is irreparable. Let go of that false thinking! Instead, solicit the insight from an uninvolved party such as a trusted friend or coach. Often, they can provide a healthy, objective perspective that can provide input on what you might do differently or better with your next attempt.

3. Once you identify the errors or causes of your sub-optimal performance, then rapidly develop and implement a course correction.

Forward momentum diminishes the pain of the failure by re-instilling hope.

4. Counter your own negative self-talk with affirming statements, such as "I am capable." Or, "I just discovered a solution that did not work. I am that much closer now to finding a solution."

5. Learn to see the humor in the situation. As in the previous chapter on humor, laughter relieves stress and puts us in a restful mood where we are apt to make better decisions. It also connects us with others, an important ingredient in overcoming any setback.

Strong leaders will push teams beyond what they believe they are capable of. The best leaders learn to accept that some failure is inevitable. If you are a leader, you will find value in applying the Fail Fast method to your team endeavors. Assess, plan, course correct and keep encouraging your team onward to success.

ENCOURAGEMENT

Besides the culture shock of suddenly being yelled at and on alert 24/7, we faced various physically demanding scenarios as New Cadets. And physical fitness at West Point was equated with leadership ability. Many women cadets were not prepared for the physical demands of Cadet Basic Training. This included the physical training of running, forced foot marches and bayonet combat drills.

Beginning by 7 a.m. we were assembled in PT uniform. We would run in formation to a location to then conduct grass drills and various calisthenics, all in coordination with a cadet barking out the next sequence.

The challenge in running at the Academy was doing so in formation. This means maintaining uniformity with the person next to you and the person in front of you. We also never knew how long we might run or what the pace might be. Often the speed was determined by the upperclassman calling cadence.

Those who "fell out" of runs were ostracized, and if a woman fell out, this was used as another reason that "they" should not be at West Point.

Based on our initial runs, cadets were assigned to three levels of running ability. Using the colors of the Academy, they were Black (fastest), Gray (medium fast) and Gold (slowest). As I was a long-distance runner, I earned a place in the Black running group within my company.

However, even with my natural ability, I was challenged to run in formation. When the leader would increase the pace, it created an accordion

Cadets running in formation Beast Barracks Summer of 1979.
Photo courtesy of Jan Tiede Swicord, USMA '83

effect on the rest of the platoon. Those of us behind the front of the group would race to catch up to those in front, all the time staying in line with the person on the right.

On one morning, when we were well into the run, the cadence suddenly picked up significantly. I estimate that we were catching up at a sub-six-minute per mile pace. This was one of the few times I wondered if I could complete the race. My chest was heaving, my legs were burning, and my mind was saying, "You are not going to make it, stop!" Just when I thought I would have to fall out, a male classmate, Ray Royalty, who was running next to me, simply said, "No, you can do it!"

At that moment, that was all I needed. I just needed a word of encouragement to dig deeper and keep going. My male classmate, Ray, risked being chewed out for talking during the run and maybe even chided for appearing to help a woman classmate. There was more to Ray than what appeared. He had begun West Point a couple of years before with the Class of 1981. He completed Beast Barracks and two weeks of the academic year, but then decided to pursue a pre-law degree at a university. In time, he went through the process of re-applying to West Point and joined the Class of 1983. When he then returned, he chose to repeat Cadet Basic Training. Perhaps because he knew the ropes, he felt compelled to assist those of us who had little knowledge of what exactly we had gotten into when entering West Point.

In my case, it was a simple declaration, "You can do it," that convinced me I could complete that run when everything from my lungs to my heart to my legs was screaming, "Stop!"

Today, some thirty-plus years later, my classmate is now Major General Royalty. His understanding of the power of encouragement surely assisted him in reaching the top echelons of the Army.

His understanding of the power of encouragement surely assisted him in reaching the top echelons of the Army.

Though I had some negative experiences with the professors at the Academy, Major Barney Forsythe was the exception. From the moment he entered a classroom, we could sense his enthusiasm for his students and the subject. Major Forsythe would break with protocols and refer to us by our first names, which was like manna to me. Finding him warm and encouraging, I set up an appointment to gain assistance in managing my time, etc. He then offered to mentor me, and we planned several follow-up meetings.

At one of those sessions, I griped about the difficulties of being a woman at West Point. And at one point, he said, "I think the best thing that has ever happened to West Point is the addition of women. It forces these men to mature so they don't end up making fools of themselves when they graduate. I have seen male cadets graduate after four years of living like monks and destroy their lives with crazy behavior."

At that time, to hear a professor endorse the inclusion of women so convincingly encouraged me to the core. There existed in leadership someone in my corner. I began to believe I could overcome the obstacles facing me.

To hear a professor endorse the inclusion of women
so convincingly encouraged me to the core.

When I bemoaned my grades, my mentor would tell me that his grade point average as a Plebe was also dismal. And after a slow start, he improved and eventually graduated in the top third of his class and USMA offered him the ability to return and teach at the Academy. Because he taught behavioral science and leadership (BS&L), I had a natural affinity for every concept presented. I was engaged in his class and I excelled.

Through his positive leadership, he helped me put my situation into perspective and encouraged me to keep at the work and to know that, in time, I would be successful.

This was perhaps the first time I experienced a leader demonstrating encouragement and then linking it to an essential leadership skill. Major Forsythe was a leader who listened and encouraged the heart of his subordinates. His example compelled me to rise above my circumstances and learn to adapt to the situations I faced. It comes as no surprise to me that he retired as a brigadier general.

The power of encouragement can never be underestimated. And this is exactly what the West Point experience can cultivate.

On the other hand, I have also worked with leaders who seem incapable of encouragement. Instead, their means of motivation were punitive and bordered on harassment.

Shortly after graduation, I reported to Airborne School in Fort Benning, Georgia. I had married right after graduation, went on a heavenly honeymoon, and then found myself in the "hell" of Fort Benning, in the heat and humidity of the summer.

As second lieutenants, we were put in charge of a "stick" or squad of Airborne candidates. With that designation, we were held responsible for all of our stick's infractions.

The three-week Airborne School consisted of one week of ground school where we learned how to land properly, one week of jumping from a tower, and then one week of actually jumping out of an airplane.

In the early 1980s, having women attend Airborne School was still somewhat new, and the Army was still learning how to integrate us. The Army feared our equipment and harnesses would catch in our hair. So, all women were required to wear a knee-high pantyhose stocking over our heads and apply a highly adhesive tape across the back of our necks. We wore the pantyhose under a heavy helmet all day long. Although we began the day before sun up, typically by early afternoon, the women had splitting headaches. The constant constriction of the knee-high hose made for a maddening sort of slow pain across our foreheads. After each day, we would pull the tape from our necks and grimace as all our neck hair was pulled from our skin.

One afternoon during Tower Week, I remember trying to both manage the pain growing across my forehead while I psyched myself up for the first jump off the thirty-four foot tower. Prior to ascending the steps of terror, a jump master (or black hat) would turn us around, tell us to bend

Airborne candidate preparing to jump from the thirty-four foot tower;
Photo courtesy of Jan Tiede Swicord, USMA '83

over and then inspect our butts to ensure the harnesses were correctly fastened as to not cause injury as we jumped.

I was still mentally preparing myself as my "stick" put on their harnesses. The black hat spied me standing still, came over and screamed at me, "Turn around and bend down." I complied, thinking he was going to check my harness. Instead, he forcibly slapped me on my derriere and declared in disgust, "You ain't taking care of your stick!"

Beyond the pain and humiliation, he had blown any confidence I had to jump out of the tower. I was mentally psyched out as I made my way to the top of the edifice. When it was my turn to jump, I froze. I simply could not muster the ability to jump out of that tower. I let someone else go in front of me and then I began berating myself. "You are a leader, why can't you jump? You are supposed to be able to lead others and here you are, standing frozen."

At that moment, another West Point woman lieutenant came up to me and said, "You can do this!" These words of encouragement were exactly what I needed. I even said to myself, "If she can do it, so can I!" So, I jumped and conquered my fear right there.

What would have happened if someone had not encouraged me at that moment? I am not sure. I may have calmed down from the harassment and jumped. Still, I am so glad that she was there at that moment.

This was a defining event for me. It was when I decided that I would seek to encourage others when they had lost confidence, felt demoralized or incompetent. Without judgment, I would seek to encourage and empower rather diminish and intimidate. And, significantly, I would also seek out support whenever I found myself in need.

I would seek to encourage and empower rather diminish and intimidate.

Years later I was providing executive coaching to an exceptionally talented woman who had risen in a male dominated industry to the position of vice president. The company was grooming her for even higher levels of responsibility. She was quite aware that she had risen to a level where her actions would be scrutinized more acutely and that her ability to lead would be tested.

As we began to work together, "Jan" shared that she felt incredibly alone. She explained that when she accepted this new role her marriage had begun to crumble. She worked long hours and felt her children were paying the price. Jan and her husband's "couple friends" were drawing away from her as the marriage disintegrated. She felt she had no one in her life left to offer support and to encourage her.

Jan was also angry at herself. She berated herself for feeling stuck and was frozen in the sense that she was emotionally overwrought yet did not have the skills to move beyond the situation.

Initially, I simply began to encourage her by offering my perspective of being a single mother myself and telling her how I learned that creating a support system enabled me to keep going despite the hurt and pain in my life.

The thrust of our coaching centered on setting up an emotional support system outside of work. She truly could not mentally engage in moving forward in her new role until she first addressed the immobilizing issues confronting her. In time and with effort, Jan was able to see the necessity of a solid group of consistent, positive people to lean upon, and then and only then, was she able to able refocus her energies on the demands of her new role.

In Dr. Henry Cloud's insightful work, *The Power of the Other*, he asserts that the best kind of relationships wire us for resilience and success. The wrong types of relationships simply do not. Those positive connections will:

- Fuel us
- Grant us freedom
- Require us to be responsible
- Defang failure
- Challenge and push us
- Build structure
- Unite instead of divide, and

- Always are trustworthy[1]

Seek those relationships that do just that and see how you will be able to manage and overcome obstacles with the right people in your corner. Avoiding negative, draining relationships is equally important. As a leader, choose to listen — and encourage — and then see how the people you lead will be more committed to the organization and will achieve more than you envisioned even when they are facing their own thirty-four foot towers.

---

1   Cloud, H., & Newbern, G. (2016). *The Power of the Other: The Startling Effect Other People Have on You, from the Boardroom to the Bedroom and Beyond — and What to Do About It.*, New York: Harper Collins, p. 217.

CULTIVATING CHARACTER

The ability to encourage those you lead cannot be overstated. Often, we may not realize how powerful is a smile, a compliment or a sincere check-in with a colleague or subordinate. To be a positive, encouraging leader requires intention but often does not require much more. Often, all we need to do is take the time to listen.

To be an encourager, we also need to surround ourselves with people who are in our corner. Those are the "positive connections" that meet all the criteria Dr. Cloud suggests. We also need to continually add to our network of connections. Below, I outline some suggestions to help identify your positive connections and grow your network.

- To begin, I suggest you assess your current personal network. Are there people in your corner who show these attributes? Do these people:
  - Fuel you
  - Grant you the freedom to attempt new endeavors and learn from them
  - Require you to be responsible
  - Defang failure
  - Challenge and push you
  - Build structure
  - Unite instead of divide, and
  - Consistently display trustworthiness

- Once you can identify those positive connections, make efforts to meet with these individuals every so often and allow them to assist you with a specific issue or challenge. My encouragers are willing to make time for me. In return, I attempt to be available for them. We are mutually supportive.

- Make a concerted effort to continually expand your network. Developing new connections is a healthy means of managing both your career and life. Tim Sanders, author of *Love is the Killer App*, recounts how he propelled his career by intentionally growing his network. He suggests that your net worth will grow in direct relationship to how you cultivate your network. He advocates growing your network by:
    - Selflessly sharing your knowledge and skills with others
    - Becoming a connector by freely and deliberately connecting your network of contacts with others, and
    - Showing compassion in all your dealings.[2]

By consistently doing these three intentional actions, people will naturally respond by sharing their knowledge and network with you. And by demonstrating genuine kindness, people will respond in kind.

Early in my career, I learned to live encouragement as I developed positive associations. I am amazed by how far a positive connection can lead me and how much a connection has helped me overcome challenging scenarios. So, I recommend that you cultivate your current relationships and form new ones by sharing your knowledge and skills with compassion.

2   Sanders, T. (2002). *Love is the Killer App: How to Win Business and Influence Friends*. New York: Random House Audio, p.3.

FEEDBACK

As I have described, cadets receive feedback from the moment they report on R-Day, which is Reception Day for our seven weeks of Cadet Basic Training. As New Cadets, we are bombarded with instruction followed by correction.

From the moment we approach the Cadet in the Red Sash, we are instructed on how to salute, how to address upperclassmen, and how to respond to commands when drilling.

Early on R-Day, I recall having to report to the cadet first sergeant of 8th company, where I was assigned. Having been yelled at numerous times up to this point for what seemed like minor infractions, I was worried about

New Cadet reporting to the Cadet in a Red Sash Summer 1976.
Source: U.S. Military Academy Library Archives Collection

saluting this large, menacing man. We were told to report using a long sentence of words in a specific sequence. As I waited behind a couple of classmates, I kept repeating the phrase in my head again and again. It was incredibly difficult to concentrate as I heard the first sergeant bellow at the New Cadets in front of me and order them to return to the back of the line. And then it was my turn. I rambled off my lines with precision, except for the fact that I began with "Sir, New Cadet Sara Fotsch reports …"

He immediately interrupted me and screamed, "New Cadet, I am not interested in your first name!" I was embarrassed and then worried. My mind raced, "Did he think I wanted him to know my first name? Did he think I was flirting? Oh no, I could be in trouble for that …" As I made my way back to the end of the line, I was belittling myself for saying something so stupid. I was on the verge of tears realizing that I had failed. I am not sure why I had inserted my first name. All my life I had always been called "Sara." I disliked being referred to by my last name. My Swiss last name was difficult to pronounce correctly, and so I preferred my first name. On another level, I believed that being stripped of my first name was one more confirmation that my previous life was being ripped away.

Somehow, by the time I emerged again to report to the first sergeant and repeat my introduction, I was composed enough to report. As I recall, I again did not report correctly, but with one further attempt I was allowed to find my room and meet my roommate. The experience of instruction followed by feedback, often delivered by harsh means, was the typical experience that first year at West Point.

As we moved into the academic year, we received constant and public feedback on our scholastic achievement. After a major exam, referred to as Written Partial Reviews (WPRs), our last names and grades would be posted outside a classroom. The academic entrance requirements at

West Point are quite rigorous with the majority of cadets being in the top 10 percent of their high school graduating classes. We were achievers and driven. Naturally, we found ourselves competing with exceptionally smart and talented peers. For many of us, getting our first bad grades was a huge blow to our self-confidence. Given that those grades were made public was further humiliation.

We were achievers and driven. Naturally, we found ourselves competing with exceptionally smart and talented peers.

Those who performed well and achieved dean's list were issued stars that they wore on their uniforms. Others, like me, who went to summer school bought STAP (Summer Term Academic Program) stars. We wore them inside our uniforms, evidence we were on the dean's *other* list.

As upperclassmen we were given leadership roles and received feedback on our performance. As a Junior (or Cow), I was a Beast Squad leader for a summer detail. At the end of the detail I received a performance review from the Cadet Platoon leader. He gave all his leaders the same score because he wanted everyone to do well. This was not helpful, but I was a bit relieved as I was both a woman and more junior in rank, and as I was merely a third-year cadet. As a result, I was not sure if I would be rated more severely. However, this leader chose to take the "cooperate and graduate" approach to his assessments, and no one gained new information on how to improve their leadership skills.

One of the more blistering critiques I endured was at Airborne School in the searing heat of summer

To view what the U.S. Army Airborne School experience entails, go to:

https://www.youtube.com/watch?v=XQJerLpLuyM

at Fort Benning, Georgia. As previously described, this Army school consists of three weeks of training. The first week was called Ground Week where we learned, practiced and perfected a PLF (parachute landing fall). We practiced for hours in sawdust-filled pits landing first with our feet, then hips and then shoulders hitting the ground. This idea was to acquire the skills to land safely and to avoid injury when we actually jumped from an airplane. The week ended with a twelve-mile run in combat boots.

If we passed Ground Week, we entered Tower Week. During this week, after securing a harness looped through our legs and around our hips, we would mount a thirty-four foot tower; when commanded, we would jump out of the tower in the correct position – feet and knees together. If successful, we then practiced maneuvering our parachutes from a 250-foot tower. If we passed Tower Week, we moved onto Jump Week, where we practiced all the cumulative skills by jumping from a C-130 or a C-17 aircraft. Then, after completing five jumps, including one executed at night (and living), we were awarded the Airborne tab and added it to our uniforms.

Brutal but deliberate feedback was the Airborne way!

Returning to Tower Week: The harnesses and Airborne parachutes were not designed to be comfortable, especially for petite women. We women developed scrapes and bruises on our necks and arms from where the harness wrenched our smaller frames. After each jump from the thirty-four foot tower, we would quickly unhook from the equipment and report to a black hat (or jump master) for feedback. A woman classmate just in front of me reported to the black hat and he bellowed, "Airborne, your legs flew out, you leaned too far to the left. And you

Airborne candidate swinging on a cable from the thirty-four-foot tower.
Photo courtesy of Jan Tiede Swicord, USMA '83

look like a Mack truck hit you!" Brutal but deliberate feedback was the Airborne way!

As a lieutenant serving in Germany, I had a great deal of responsibility with units of more than 300 personnel. After a training exercise, we conducted after action reviews (AARs). All the company leaders, both the officers and the enlisted non-commissioned officers (NCOs) would meet and discuss in detail the exercise. We would review what went well and what we could do better next time. This real-time learning then informed our planning for the next exercise, where we could improve our performance. In this sense, the Army is a great learning organization.

We received Officer Evaluation Reports (OERs) annually. Typically, we would write our assessment for our commanders. Many of them simply rubber-stamped what we had penned. Fortunately, in other cases, my commanders would write superlatives about my performance. Overall, the feedback I was given as an officer was never harsh or demeaning.

Depending on my superior, the feedback was either mostly positive or so standard that I never learned much.

It may be that during the formidable years at West Point when I perceived more negative than positive critiques, I became overly sensitive about feedback in general.

When I began to work in the civilian world, I was exposed to cultures that offered fewer opportunities to receive feedback. Working for busy managers most of my career meant I rarely got ongoing feedback like my experience in the Army. Often, I would be surprised with some of the comments on my annual performance appraisal.

As I moved into consulting, I would solicit feedback from a client after each event or project. My tendency, however, was to become defensive with feedback that I deemed inaccurate or unfair. At times this defensiveness resulted in losing influence, damaging a relationship or even losing a customer.

When I was in the midst of my master's degree program, I was exposed to organizational psychologists Robert Kegan and Lisa Lahey's *Immunity to Change* methodology.[1] This duo discovered that we all have "hidden competing assumptions" that drive unhealthy actions.

To learn more about the *Immunity to Change* methodology and discover your own competing assumptions, I highly recommend their book:

Immunity to Change: How to Overcome It and Unlock the Potential in Yourself and Your Organization.

---

1   Kegan, R., & Lahey, L. L., & K. (2009). *Immunity to Change: How to Overcome It and Unlock the Potential in Yourself and Your Organization.* Boston: Harvard Business Review Press.

In my case, I desired to grow my consulting practice and become more successful. To do this, I needed positive referrals from very satisfied clients. To gain these referrals, I needed to continue to improve and get better at my consulting work. One way to improve was to ask for feedback from my customers, receive it and then make the necessary improvements.

However, my response to criticism or negative feedback seemed to undermine my overall business objective. I tended to become defensive or deflect the input that I might receive. If truth be told, a client may have stopped providing input based on my earlier responses. At times this tendency put me at odds with a client, and I saw my ability to influence him or her ebb away.

What Kegan and Lahey's approach helped me unearth was the hidden competing assumption that I was equally committed to upholding. By hidden, I mean it was unconscious, something I was not yet aware I was doing. Through the duo's deep questioning exercise, I discovered the following competing commitment:

> I assumed if my clients pointed out errors or suggested better ideas than mine, my value in their eyes would be diminished and they would no longer need my services.

I defended and decried the very input that I needed to perform better.

Hence, I believed my value was derived from *their* perceptions and I was not willing to have that jeopardized. I defended and decried the very input that I needed to perform better.

The next step Kegan and Lahey propose is to "test" this unconscious assumption. Initially I did this by simply listening to feedback from a safe source; in my case, that was my husband. I listened to his input concerning a problem and realized that his solution was another option. After employing his suggestion, we solved the issue and I was still valuable in his eyes. My action had not diminished my worth.

Armed with this new evidence that cast doubt on the veracity of my competing assumption, I eventually gained the confidence to employ my new approach to accepting input with a long-term customer. I learned to listen, fully listen, and thank them for their feedback. As I listened, I discovered that their input was incredibly valuable. With their feedback, I could more easily make a course correction if I had fully understood their expectations. On consulting issues, the client may be closer to an issue than a consultant. Their perception of an approach might be the needed data point I was missing. Indeed, clients from whom I elicited input became a wealth of information. I was more informed, and hence, my solutions were more effective and more readily adopted. I was not working as hard as I had previously. Although I can still cringe a bit when I receive criticism, I have begun to view feedback as a gift rather than a death sentence.

Clients from whom I elicited input became a wealth of information. I was more informed, and hence, my solutions were more effective and more readily adopted.

In my years in consulting, I found that there are many people who, like me, do not receive feedback well. The truth is that we will continually receive input from others whether we want that information or not. We also know that those who can take in feedback and adapt their behaviors often advance more rapidly.

Now, as I reflect on the seemingly endless critiques I received at the Academy, I realize that I did not have the tools at that time to properly assimilate all those data points. Still, receiving feedback is inevitable; learning how to accept and apply what you can from that input is truly the only means to improve.

CULTIVATING CHARACTER

It has been my experience that those of us who are high achievers often struggle to process input. We may bristle, especially when barraged with feedback delivered harshly. To be able to respond appropriately and accept valid input means we have to sort through what is helpful and let go of the rest.

Self-awareness is the first key to change.

A good place to begin to learn to readily accept feedback is to reflect on how you currently handle criticism. Are you able to listen to an appraisal or do you immediately become defensive? Self-awareness is the first key to change.

Here is a formula to let a person giving feedback know that you have heard their message.

1. Repeat the feedback and ask if your understanding is complete. For example, "I hear what you are saying. When I take over in the meeting, I am overstepping my role and making it appear that I am running the project rather than you. This makes you feel disrespected. Do I understand you correctly?"

2. Next, listen. They may amplify the feedback, allowing you to gain greater clarity about the issue.

3. Assess if there is some validity in what they have mentioned. If there is validity to their feedback, thank them for sharing this information with you and assure them you will give their feedback a good

amount of thought. If it is simple for you to make this change, then agree to make the change and even ask them to remind you if you do not change.

4. However, if the change is not necessary, simply thank them for the feedback. You can then share the feedback with a couple of "positive connections" (see the chapter on Encouragement) to further discuss the validity of this perception.

As leaders, we have the responsibility to provide feedback to our employees so they can improve their performance and our organization can be successful. Here are some suggestions on how to give feedback:

- Give frequent feedback. Subordinates need to know not only when their performance is not meeting expectations but also when it is exceeding those expectations. Giving *only* negative feedback will typically demoralize; however, avoiding negative critique is not conducive to employee development and organizational productivity either. Not giving any feedback serves no one. Frequent feedback requires that you, as a leader, are engaged in your teammates' work. Walking around, asking questions and engaging with your subordinates is one means to give ongoing and timely feedback.

- Deliver the feedback by initially describing how the conversation will unfold and encouraging the employee to engage in a dialogue about the behavior. Below is an example of this type of conversation:
  - "Jane, I want to speak with you about some concerns about the project update meeting. I believe it may have caused some frustration for the customer. First, I want to review what occurred and then share my concerns. Then, I would like to hear from you to see if you also share those concerns. Once we agree on what happened, I may want to say more about my concerns. After this, we can decide what, if anything, we need to do going forward. I am open to the possibility that I may have

missed something or contributed to the issue. How does this process work for you?" [2]

- This approach provides your subordinates with the ability to buy in to how the conversation will take place and creates the opportunity for learning by both parties. [3]

- Ask for feedback from your subordinates. Asking about how you are perceived builds rather than erodes trust. It demonstrates to your charges that as a leader you also can improve. If you accept that feedback with grace, you are also modeling how your subordinates should take in feedback they might receive.

With all feedback, work to see it as a gift, a new data point of how others perceive you. Keep this opinion as just one of several perspectives. Critique is also a building block for learning humility – the next principle I will share.

2   Adapted from Schwarz, R. (2017, October 30). The "Sandwich Approach" Undermines Your Feedback. Retrieved June 14, 2018, from https://hbr. org/2013/04/the-sandwich-approach-undermin.

3   Ibid.

HUMILITY

Required lectures were part of our leadership development training at West Point. One evening after dinner, my class of nearly 1,000 cadets and I filed into Eisenhower Hall. This was one of numerous mandatory seminars we attended as cadets. Typically, these events lasted an hour or more. As a cadet, I rarely found the speakers interesting. Most often they were male generals or "Old Grads" who had returned to imbue their wisdom on often less than enthusiastic cadets.

I remember thinking, "Great, I have to listen to another old guy and I have hours of homework and a WPR tomorrow. I want to get to bed early for a change." Sleep was always on the mind of a cadet whose days began as early as 6 a.m. and often lasted until midnight or later.

Despite my negative attitude, I vividly remember one general. He was not very tall and even a bit rotund. His opening line is what has remained with me. He began with something like: "I am not a West Point graduate. I am not Infantry. I am not Armor. I am not a Ranger. I am not even Airborne. What you are looking at is a complete and utter failure."

The audience chuckled, but I understood the irony. He was not what most of his audience expected in a general, and yet, here he was speaking to us.

I was immediately engaged with this relatable and self-deprecating man. As a woman, I was barred from many of the typical ways to earn the rank of general. I was not allowed in the combat arms branches of the Army, such as the infantry, armor or special forces. Ranger School was not an option for me, although I did eventually graduate from Airborne School. After hearing him speak, I began to believe that perhaps, if this man could obtain the rank of general then anyone, even I, could reach the top echelons of the profession of arms.

I have come to believe that humility is rarely taught;
perhaps it is best caught.

This man, whose name I do not recall, exemplified what humility is all about. He was perfectly comfortable within his own skin and by so doing, he exuded an approachability that made me admire and emulate his approach. Isn't it remarkable that I remember this speaker over the many others we heard during my cadet years?

I have come to believe that humility is rarely taught; perhaps it is best caught.

Cadets are given officer sponsors at the Academy. And I was fortunate to be matched with Major Palmer Bailey, who taught in the Department of Geography and Computer Science. A brilliant man, he was later given an assignment to the Astronaut Office at NASA's Johnson Space Center in Houston. After retiring at the rank of colonel, Palmer and his wife, Bonnie, and their children moved to Alaska, where he personally built their own log cabin as well as the very road to their homestead.

At West Point, the Baileys welcomed me into their officer quarters and often went to great lengths to support me. One night I was returning to the Academy after one of our few breaks. My flight was delayed, and by the time I arrived in Newburgh, NY, I had missed the last bus to the Academy. There were a couple of other cadets similarly stranded; we were very concerned that we would not make it back before Taps. I found a phone booth and called Major Bailey and explained my situation.

True to his nature, he immediately got in his car and made the trip to pick me up.

While I was making this phone call, another cadet had somehow secured a taxi. As I got out of the phone booth, he motioned to me from the car; they were leaving immediately, and if I wanted a ride, I must get in the taxi or be left behind. Wanting to get back to USMA as quickly as possible to be sure not to be late, I did the unforgivable; I jumped into the taxi. When I arrived at the barracks, I immediately used the one company phone to call the Baileys and explain what had happened.

By this time, Palmer, as he allowed us to call him, was back home. I was surprised that he was not angry when I called to apologize: The long ride home may have given him time to cool down from the incident. He accepted my apology and understood that I would have called him back at the time had there been time. He didn't question my lame excuse and continued his support and encouragement throughout my time at the Academy. Apparently, not holding grudges is also a characteristic of the humble leader. In fact, Bonnie and Palmer hosted this intrepid cadet's bridal shower a few years later.

Not holding grudges is also a characteristic of the humble leader.

Earlier in this book, I mentioned Major Barney Forsythe, the psychology professor who became a mentor to me. What I only appreciated later was that Barney had devoted much of his personal time after hours to instruct and encourage me. It was time he could have spent with his wife and children. His willingness to invest in a cadet at his own personal expense, and to do so without expecting anything in return, was an example of humble leading. Serving others when it may be inconvenient or require sacrifice are also elements of a humble leader.

I learned about humility not just from positive examples from Palmer and Barney; I was personally humbled by the entire experience of the Academy. I soon realized that I was surrounded by many very talented and gifted people. I was not as smart as I thought I was.

I learned to seek help when I struggled academically, often from a more gifted classmate. This necessity developed some needed adjustment to my possibly inflated self-perception.

When we do our duty without seeking recognition or credit,
this is when we begin to embrace humility.

The duties given to a Fourth Classman also taught humility. Whether delivering mail or laundry, we learned to take pride in those responsibilities, doing what needed to be done with excellence. When we do our duty without seeking recognition or credit, this is when we begin to embrace humility.

This recognition extended to how we were to lead the cadets in our charges and eventually the soldiers we were to command. We were taught to think of them as "our" squad, platoon or company. Later, as leaders in the Army, we knew that we needed to always care for "our" soldiers.

One way this was manifested was during a deployment when "chow" was delivered. As officers, we made it a point to wait to obtain our meal until every last soldier had been served. In this way, if rations were short, our soldiers were not the ones impacted. This was a simple act of humility.

Athletic competition could also inculcate a humble spirit. We learned that everyone on a team contributed to the overall success of the team. Even if we were star athletes in high school, that did not mean we would

The West Point Cross Country Team, 1980. Author is bottom row, second from right;
Photo courtesy Harlene Nelson Coutteau, USMA '82

be at the top of the team at the Academy. Through competition, we would learn to think of team success over individual success.

When I was a Yearling, the women's cross-country team won the Eastern division championship. This accomplishment qualified us to attend nationals! Injury free, I had one of my best performances at the divisional competition. However, the week after the event, I found I had severely inflamed my right hip flexor ligament. When I attempted to run, I experienced a sudden sharp pain in the front of my hip. Despite my injury, our coach asked me to accompany the team to nationals in Seattle, Washington. He admonished me to rest for a week and then plan to run at the competition. So, I did not run that week. In the evening as I studied, my right leg was propped up on a trash can with an ice pack on my hip. By listening to my coach, and with rest, elevation and ice, I prayed my ligament would heal so that I would again compete with my team.

The team flew to Seattle and then lined up for the race. From my first stride, I felt a sharp pain return and was forced to limp. Because I was a part of the Army team, I was determined to complete the entire three miles even as I endured sharp spasms with each stride. I wanted to represent my team and the Academy even if I had a dismal finish. As I crossed the finish line, my teammates were waiting for me and cheered me on. They showed a humble spirit by demonstrating empathy for a teammate with physical limitations whose sub-par performance might have negatively impacted the overall team's score.

They showed a humble spirit by demonstrating empathy for a teammate with physical limitations.

I had learned humility by competing while not at my physical best. And these types of situations help us to appreciate the limitations of others as well. We can learn compassion for others, which is another characteristic of the humble.

Whether through Plebe duties, academic necessity or athletic endeavors, I was humbled. I learned to ask for help when needed and then supported the team over my individual desires. Those self-effacing leaders at the Academy impressed upon me the desire to become more like them. My education in humility continued even as a "green" lieutenant.

I learned a good deal about unpretentious leading from an outstanding warrant officer who served under me in my first unit in Germany. Mr. "Brooks" was not boastful or overbearing. Instead, he quietly went about running his platoon of highly specialized armament technicians without fanfare. He proactively ensured that his team had all they needed to do their jobs. At times, he needed to go to extreme measures to find repair parts for the antiquated equipment of a job order. He might travel hundreds of miles to an Army depot and sort through piles of inventory to find a single scarce part.

Further, Chief Brooks always found ways to motivate his team, especially on our six to eight weeks of deployment. His repairmen often worked around the clock for several days fixing critical armaments for our Mechanized Infantry Division customers. So, Mr. Brooks cleverly devised a means to reward and energize his platoon. First, he "secured" a refrigerator. I never asked how, and I didn't want to know. He then camouflaged it with olive green and black paint. Then, he loaded it onto the back of one of his trucks that was equipped with electrical power. Before each exercise, the fridge would be stocked with plenty of candy bars and sodas. He would charge the soldiers only enough to replenish the store for the next deployment. His self-effacing leadership style proved highly effective. There were rarely any rejections on the unit's repair work and were very few or no disciplinary issues in his armament platoon. As a result of Chief Brooks's non-assuming leadership, he earned the respect of his team and his superiors.

Several years later, I was new to Richmond, Virginia, and in need of a corporate position. I had successfully secured an interview with a very competitive corporation that proclaimed, "We only hire the best and brightest." Its interview process had the reputation of being unpredictable and brutal, requiring logic and mathematical skills. These were skills I had not used in years!

I reached out to the West Point Society of Richmond and was connected to Kevin Beam, USMA '78. I mentioned I was preparing for a critical interview and needed help. Kevin volunteered his assistance and even mentioned that he had also interviewed with this company. What I did not know was that Kevin was a brilliant mathematician. After graduating from West Point, he went on to obtain a PhD in operations research and statistics. He had also served as a math professor at the Academy.

Kevin expertly recalled in detail the types of questions and assessments that I would be given. And he patiently taught me algebraic methods I had long forgotten. He mentioned that there was a key moment in the interview when I should stand and present my solution. All told, he spent several hours with me, prepping me for the interview. Grateful for his assistance, I listened, practiced and applied what he shared. I got the job! This was largely because a gifted, yet non-assuming man was willing to freely offer his assistance. Finally, another element of humility is a generous spirit, one willing to share and teach his or her knowledge.

Another element of humility is a generous spirit, one willing to share and teach his or her knowledge.

Later, as a consultant, I would tackle progressively more difficult problems for clients. It became very natural for me to find and add people to my team who were more knowledgeable or experienced. I would ask them for their help and assistance. I believe this is why my projects and teams were successful. I sought out people who would make the overall deliverable better than if I had done all the work myself. In some cases, I made a great deal less personal profit on projects, as I made it a point to amply compensate my contributors. Paying my people well paid off because often the client would hire my firm again simply because the deliverable was excellent.

Humility begins with putting the needs of others above our own. It may require doing the uncomfortable or the inconvenient. It also requires you to defer the need to be in the spotlight. With self-effacing leadership, you accept that you do not need to be the smartest person in the room. Instead, you learn to be comfortable with the gifts you *do* possess while recognizing the talents and contributions of others.

CULTIVATING CHARACTER

If I asked you to name the most influential leaders that you admire, whose names might appear on that list?

Now, if I were to ask you to name as many humble leaders as you can, would that list be much more abbreviated than the first?

That is the quandary we often find when trying to identify role models exhibiting humble, unpretentious leadership. We can think of many influential leaders, but many of them would not be characterized as being humble. In fact, in the Western world, we are often drawn to a leader's physical appearance and charisma over humble attributes. Perhaps because we are enamored by the beautiful and engaging, we miss those in our midst who are demonstrating leadership through humility. However, it may be the humble leaders who have the most long-lasting impact.

If you want to develop a self-effacing leadership style, here are some suggestions:

- It is not about you! Embrace this statement and place your focus on others.

- Intentionally and consistently put others before yourself. Modeling a team focus encourages others to choose to follow you. Begin by committing to making more "we" statements than "I" statements.

- Invest in understanding your strengths and weaknesses. Sharing both with your team builds transparency and trust. Then seek to understand each team member's strengths and weaknesses. Building this level of vulnerability will allow you to leverage your team's strengths more effectively.

- Resist the urge to handle everything yourself. Ask for help and input and ask yourself whether the end product is better than expected.

- As a leader, set a vision for your team or department and work with your team members to set collective team goals – objectives that they must rely upon one another to achieve. Once those goals are achieved, celebrate the team's success. Keep the focus on the team and not you.

I once heard an introduction for a speaker done in an unusual manner. As expected, the presenter initially highlighted some of the speaker's remarkable achievements. However, what the announcer said next is what captivated me. "His many achievements were not the reason I asked him to speak. What I truly find remarkable about this man is that after being around him, I am a better person and I want to be even better."

"What I truly find remarkable about this man is that after being around him, I am a better person and I want to be even better."

That is a humble leader, I thought, one who by their very nature makes us better. Unfortunately, those leaders are rare, and more often than not, we will instead have to work with and for those who infuriate and aggravate us. Read on!

PART III
Navigating the Forces

The next four chapters describe leadership skills that are acquired when living within a system that seemed intent on keeping me from advancing. You will walk in my shoes through situations that are both positive and negative. In retelling some of these incidents, I have not shielded you from some of the more negative experiences I encountered at West Point. I firmly believe, however, that you will discover that the lessons learned in the crucible did not break me; instead they became foundational for how I have navigated other difficult life experiences. My intent is for you to gain insight in how you might learn to navigate similar scenarios in your own life.

THOSE DAMN WOMEN!

You cannot convey the West Point experience if you do not tell some stories about the women at the Academy. Although the vast majority of my interactions with my women classmates and women senior to me at West Point were positive, there were several interactions with upper-class women that were incredibly negative and disappointing.

In fact, there were a couple of women who at the time, demonstrated how *not* to lead. One such woman happened to be on the cross country (XC) running team where I competed. Sometime during our seven-week Cadet Basic Training, we switched from wearing our running shoes to running in basketball sneakers. Sneakers are inadequate shoes for running on pavement. Add to this the back and forth acceleration and deceleration while running in formation and I began to feel a dull pain in my lower leg. When the academic year began, our XC training ramped up with longer, more intense runs, and my pain was exacerbated. After a visit to Keller Army Hospital, I was diagnosed with a stress fracture and given a medical profile.

As a Fourth Classman, my medical profile stated that I no longer had to "ping" (or walk briskly at 120 steps a minute). Plebes also squared corners when entering a room or stairwell. As you can imagine, pinging added many more steps to a Fourth Class cadet's day and would further aggravate a stress fracture. A Fourth Classman's uniform is noticeably short of insignia, so I was easily identified. Further, for a Plebe to not be pinging from place to place, any upperclassmen in my path might stop me and demand to know why I was not "moving out." I was inevitably stopped and questioned.

Once stopped, the upperclassman would often find fault with my uniform or my Plebe knowledge, etc. When questioned, Plebes were only allowed to reply with one of four statements:

1. Yes, Sir or Ma'am

2. No, Sir or Ma'am

3. No excuse, Sir or Ma'am, or

4. Sir or Ma'am, may I make a statement?

A Plebe could get flustered, especially when two or more cadets were interrogating and demanding recitations of Plebe poop. You also never wanted to call a ma'am, sir, or visa-versa as this would only bring more hazing. With my medical profile, when halted, I would ask to make a statement and then state, "Sir (or Ma'am), I have a medical profile."

One afternoon, I was walking (actually limping) across an area between barracks when my XC teammate stopped me and yelled, "Halt! Cadet Fotsch, why are you not moving out?" I knew she was well aware of my stress fracture. Every day at practice she saw me heading to the pool to work out rather than run. I could not believe she had stopped me. I was hurt and then became angry. The incident made me wonder whether she cared about her teammates.

I came to learn that Cadet "Over the Top" was an extreme rule-follower. It was rumored that she even wrote up her own roommate for having dust bunnies under her bed.

As varsity athletes, we were assigned to Corps Squad tables in the Mess Hall. Ever conscious of our weight, our running team elected to be on "diet tables," which meant that we did not get dessert, had more salad at

Female cadets at a Corps Squad Athletic table.
Source: U.S. Military Academy Library Archives Collection

our tables, and drank water rather than milk. Sitting with our teammates was a benefit; we Fourth Classmen experienced less hazing at our team tables. However, we still completed Fourth Class duties such as memorizing the beverage preferences of the upperclassmen. Our XC team was too large for one table, so we shared half of our table with part of the all-male pistol team.

The men on this team were all upperclassmen. They began to call the women Plebes by our first names, which was prohibited by Academy regulations. Upperclassmen could not acknowledge the first name of a Plebe until Recognition Day, which typically is held just before graduation in May. As Plebe women, we always responded appropriately to these men with "Yes, Sir," and "No, Sir," etc.

Somehow "Top" got word of the men's infraction and wrote the entire table up; each of us received demerits. Word came down that Cadet Over the Top thought her Plebe teammates were in the wrong. We were at fault for not reporting the pistol team's violation of regulations to her!

As Plebes, we feared retaliation when reporting a violation made by upper-class male cadets. And as women cadets, we had not been coached on how to handle a situation like we experienced. All I knew was that upper-class male cadets were off limits, and I was fully willing to abide by that regulation.

As Plebes, we feared retaliation when reporting a violation made by upper-class male cadets.

At that table, we were the subordinates; the men of the pistol team had power and authority over us. In that sense, they should have been disciplined more severely (and they may have) while we might have been given instruction on how to handle situations like that in the future. Punishing all did not consider that the Plebe women at the table may have been the victims of unwanted attention and recognition. Instead, it was assumed that we welcomed these advances. A better leader might have interviewed us prior to writing us up and letting our tactical officers (TAC) mete out the appropriate discipline based on the facts. As it was, my TAC never inquired about the incident. He assumed total guilt and assigned several demerits. Nothing was gained in terms of how I might handle a compromising situation in the future.

To be fair to Top, she was a young and inexperienced leader who may have truly believed she was doing her duty. She may have thought she needed to protect the reputation of the XC team. However, her approach eroded trust and angered her teammates.

The problem I have with leadership decisions like this situation demonstrates is that it is played out all the time in sexual harassment complaints in the working world; the *#metoo* movement is predicated on this notion. Too often, leaders fail to fully investigate. Sometimes a person (male or female) is assumed to be guilty or partly at fault for simply being in a situation. Without considering the very real boss-subordinate dynamic, the victim is often presumed to be guilty or partly at fault and left to defend him or herself. The pistol team scenario at West Point stung all the more because it was a woman leader who instigated the punishment, and we were never given the opportunity to share our side of the story. Sadly, she was a woman leader who I had hoped to learn from and to emulate.

The pistol team scenario at West Point stung all the more because it was a woman leader who instigated the punishment.

On balance, though, I found that most of the women who were senior to me were supportive. The only problem was that we knew so few of them. The women at West Point numbered just one for every ten men. The Class of 1980 began with 119 women, of which only sixty-two graduated (a 52% graduation rate), and we were spread throughout the thirty-six companies of the Corps. There was little effort to get us together and some of the efforts that were made felt forced and unproductive.

I found that there were other women on the cross country and track teams who were effective role models and encouragers. Many were doing well academically, advancing in leadership roles and were approachable and empathetic. They were strong, effective women who did not seem to have an axe to grind. I am thankful for the likes of Sue, Meg, Gail, Harlene and Roberta, to name a few. They came to be my friends and mentors.

When I was a young lieutenant stationed in Germany, I had the misfortune of working for another woman who used intimidation as her leadership operandi. As our company commander, the captain called a "vitally important" meeting for her staff for a Saturday morning. Since we typically worked fourteen-hour days, five days a week, and had extended six-to-eight-week field exercises where we worked seven days a week, we relished our limited time off with our families. I arrived along with the rest of her leadership staff in our battle dress uniforms (BDUs) at 0700 hours or 7 a.m., as instructed. We then waited until 9 a.m. for the captain to make an appearance. When she showed up, she never apologized for being late, and to make matters worse, in complete disregard for Army protocol, she showed up in civilian clothes. The captain spoke about an upcoming exercise for about 30 minutes and then dismissed us.

My in-laws had flown in for a visit; because my commander delayed the meeting until it was convenient for her, I missed joining them for a cruise on the Rhine River. This leader liked to remind us that she had us "under her thumb" and could call us in whenever she chose.

Later that week, there was a battalion social and I brought my in-laws to the event. My mother-in-law, Paula, was the quintessential Southern lady; beautiful, charming, yet able to get her point across in a subtle way. When I introduced my in-laws to Commander "Thumbelina," Paula took her hand, looked into her eyes and said something like, "It's such a shame that Sara could not spend her Saturday with us on the Rhine. You know, we came such a long way and we are not here for very long. Sara should really get some more time off."

At the time, part of me cringed and worried Captain Thumb would use this request against me. At the same time, I thoroughly enjoyed seeing Thumbelina's facial expressions change as she listened to Paula. Thumbelina squirmed to release her hand from Paula's and stumbled in her reply, stating that an upcoming exercise would prevent any time off (even though that exercise was three months out). Undeterred, my mother-in-law reiterated her point, "That is too bad; you do realize that we are only here for a short time." Paula was going to make sure Thumb knew she was in the wrong.

Captain Thumbelina was also not physically fit. She refused to partake in physical training (PT) with her company. Instead, she would run on her own with her husband, who was also an officer. As you can imagine, this went over like a lead balloon with me. After coming up against so much animosity for women who could not meet the physical requirements at West Point, I had very little empathy for a woman officer who would not lead PT with her company. For me, she was making it more difficult for women by not leading by example.

With the perspective of time, I realize now that Thumb lacked confidence in herself; this led her to try to control her subordinates into submission. One of her techniques was to give a directive (an order in military speak) to members of my platoon and not let me know she had done so. This left me suddenly accountable for some action that I had no idea needed to be accomplished. Perhaps you also have had a leader whose actions undermined your authority?

Thumbelina used this tactic with all her platoon leaders to keep us off balance. At times, I felt she did this so that she could find some reason to find us at fault. As you can imagine, this made it more difficult for us to lead our soldiers. Despite several efforts on my part to request that she inform me first before directing my troops, I soon realized that my words were falling on deaf ears. She seemed unwilling to trust her platoon leaders; instead she sowed discord within her company.

Exasperated, I finally went to the battalion commander and explained that I could not work for Captain Thumb. She was undermining my ability to lead my platoon. He listened and agreed that her actions were not professional. To my surprise, he allowed me to take a role on his staff as the assistant S2/S3. In military parlance, an S2 is the intelligence officer while the S3 is the training and operations officer. In our Forward Support Battalion, the role was combined into the S2/S3 Office. As the S2/S3 assistant, I did whatever the S2/S3 did not have the time or interest in doing. Hence, I spent long hours on projects such as compiling operations-related assessments and writing reports.

While I had a new job, another woman lieutenant remained in the company. When LT "Smith" also got fed up with Thumb's treatment, she went to the battalion commander and asked to be removed from under this weak leader. Unfortunately, Smith was told that the company could not lose another LT, and she was forced to stay under this dismal captain. Smith suffered with a less-than-stellar Officer Evaluation Report (OER) from Thumbelina despite being a highly competent junior leader. In time, Captain Thumb was relieved of duty and left the Army. Unfortunately, she left in her wake a bad impression of other female company commanders.

I have also struggled as a woman in leadership.
There were times when I was aggressive, rather
than assertive, and militant, rather than diplomatic.

To be candid, I have also struggled as a woman in leadership. There were times when I was aggressive rather than assertive, and militant rather than diplomatic. As a young lieutenant, my battalion deployed for several weeks to a remote training site. We were housed in several nondescript buildings. The S4 (or supply officer), had secured a large bay for the male officers. However, he had made no such arrangements for the women officers. We were simply told to find a cot with the women enlisted soldiers.

I was irritated and thought, why were we not housed with our male colleagues? This decision removed the few women officers at the site from contact with our superiors. I was convinced that the decision was inequitable and dismissive. Enraged, I charged into the S4's makeshift office and demanded to know why the women officers were not billeted with our male counterparts. He tried to explain that he did not have another building to offer us and he assumed that we would prefer to stay with other women. I argued that we should be billeted with our colleagues. I suggested it was easy to use sheets as dividers to ensure our privacy. A bit blown-over by my brazenness, the S4 acceded, and all four women officers moved to the male officer building.

However, what I did not foresee was that most of the male officer cadre snored! We are talking about more than twenty men snoring like chainsaws all night long. After one sleepless night, all four women returned to our original quarters. My tirade with the S4, however, did not yield changes in billeting for the next deployment. Instead, because I lacked diplomacy, the S4 avoided me and our working relationship suffered. I had not mastered the dragon lady within.

Later, in my corporate career, I worked as an internal consultant for a very large company. A woman senior vice president, "Mad" had the reputation for losing her cool and publicly disparaging her staff. Often, people did not understand what would prompt her outbursts. She intimidated most people because her terror rages were unpredictable.

One afternoon, she called me into her office to ask that I conduct a 360° assessment on one of her managers. The 360° assessment is an electronic method of gaining feedback on the perception of a person's abilities from the person's boss, co-workers and employees, thereby providing a 360° view of their performance. Through questioning, I learned that the SVP was interested in uncovering this manager's weaknesses to build a case to end her employment.

As a professional coach, I knew this was never the intended use of a 360°
assessment, and I was feeling pressured to divulge anything unfavorable
about this manager to this menacing woman. I met with the manager and
asked her why she thought the SVP was asking for this assessment. She
freely disclosed that she suspected Mad was planning her demise.

Aware of this, I decided that regardless of what was revealed on the
360° assessment, I would not divulge it to the SVP. I would assert that
the tool's feedback is provided directly to the individual. It would be
the employee's decision to reveal those findings to her boss or superior.
I knew that in choosing this approach Mad might criticize me or even
seek my own termination. However, I would not be her puppet.

I conducted the 360° and it revealed that the manager was well regarded
by her immediate boss, team and co-workers. By all measures, this woman
was perceived as conscientious, competent and effective. When the SVP
asked for a summary, I simply stated that her assessment indicated she
was competent and could be considered for advancement. I remember
that the SVP stared blankly at me for a few seconds and moved on to
another subject. I believe that once she realized she could not get the
ammunition she desired from me, she dropped the subject entirely.

It may be that I stood my ground with this woman and did so with some
measure of grace because I had practiced standing up for myself at West
Point and as a young officer. Not all those attempts were successful. But,
the point is, I had developed my confrontational skill set with practice.

Over the years, I have had several women bosses and clients. Some have
been competent, excellent leaders and comfortable in their own skin.
In other cases, I have seen women use their positional power to control,
manipulate and intimidate. I have experienced woman-on-woman jealousy and a boss who disparaged my work simply out of envy for more of
the spotlight.

I have had enough questionable experiences with women leaders to want to ask, "What is it with those damn women?"

Part of the answer comes from several studies that have been conducted comparing men and women leaders. A recent study amassed 360° feedback data on about 16,000 leaders, covering a wide variety of industries. Of the data set, two-thirds were male and one-third were female. The conclusions include:

- Women leaders believe they must be more than perfect to succeed. The drive for perfection may make them appear as both demanding and aloof.

- Women still provide the majority of the childcare planning arrangements. Even as they seek to move up in a career, this added demand can create extra stress, which can come out in inappropriate attitudes and behaviors.

- Women and men start out with little perceived difference in their skills. Early on, however, men are viewed as better leaders. Yet over time, if the women survive and advance through to higher echelons, they tend to be perceived as more capable leaders than men. This improvement is linked to the readiness of women to ask for and receive feedback and their commitment to self-improvement.[1]

At one point, my own perfectionist drive had an unhealthy outcome. Women having children while in the Army is incredibly challenging given our deployment schedules and the possibility of being called into combat. I also knew that my male superiors might view having a child as a liability and it could negatively impact my career. Armed with this

---

1   Bob Sherwin, Contributor. (2014, January 24). Why Women Are More Effective Leaders Than Men. Retrieved March 15, 2018, from http://www.businessinsider.com/study-women-are-better-leaders-2014-1.

mental construct, I was determined to prove that having a baby would not impact my performance in the least.

Shortly after my first daughter, Larisa, was born, my annual PT test was due. During my maternity leave, I walked, swam and ran, all while breast-feeding and caring for my newborn around the clock. When it came time for the test, I excelled. However, later that day I also fainted. After speaking to my doctor, he reminded me that I had just had a baby and it would take a year for my body to return to normal. I had demanded my body to do so after only six weeks. Like many women, my drive to prove myself worthy added undue stress and my body recoiled.

Women leaders are working against long-held beliefs and bias in the work place. To be viewed as perfect, many women are under tremendous amounts of self-imposed stress. It is no wonder then that we may become defensive or aggressive. And for some of us, we may tend to view other women as competition for that rare seat at the table. Being mindful of the unique circumstances facing women leaders informs our response.

It appears, though, that we women will accept feedback more readily than our male peers, so there is plenty of potential for our continuing maturity and leadership capability.

CULTIVATING CHARACTER

Maybe you find yourself working for one of those "damn" women and are struggling with how to complete a project or a job or simply survive her wrath. Begin with the belief that this situation is not unique. There are many who have dealt with difficult, demanding women managers, clients and co-workers. You may need to try several methods to manage an overly demanding or aggressive woman.

Alternatively, you may be a woman facing inequity, bias and/or multiple demands and expectations of yourself. How do you assert yourself and not become labeled as unhinged?

Here are some techniques I have employed to address these two scenarios:

- Learn to manage your emotions. As I have mentioned earlier, when you master your emotions, you are better able to think and engage another person in dialogue.

- Become aware of your body language. Learn techniques to remain calm. One technique is to place your hands on your knees with your palms facing upward and lean toward that person. It is nearly impossible to become defensive when in this position.

- Listen to the person's demands and then quietly repeat the expectations. When you model control and soften your tone, most people respond by lowering their weapons, so to speak. They will slow their speech and demonstrate less agitation.

- Ask questions only to clarify the demands. Often, when you remain calm and repeat unrealistic demands, a person may realize they are being unrealistic.

- In a calm tone, state your concerns, and relate your objections to the overall team goals.

- If you are still pressured to do something that you do not believe is within your scope, is unrealistic or is not aligning with your ethics, then state your objections while remaining calm. In so doing, be willing to accept the consequences: always protect your peace of mind and emotional well-being.

After reading these pages, you might see that you have some of those traits of the abominable and you are wondering what you might do to repair your public persona.

As a woman and a leader, I must continually evaluate if, in fact, I am becoming one of those "damn women." Those who have experienced my critical eye may feel like I am looking for errors in work products rather than looking for the good. I hope that the humbling experiences (of which I have had nearly as many as the triumphs in my life) have kept me from the extremes.

Regardless, I know I have had bad days when the pressures of life have overcome me.

Learning to quickly apologize helps restore trust. Deliberately requesting feedback from clients, colleagues and subordinates also keeps me from becoming a dragon woman. Finally, learning to laugh at my foibles keeps me sane. These techniques are also useful when dealing with the opposite sex, which is the subject of the next chapter.

ABOUT THOSE MEN

There has been much written about the chauvinistic, overbearing, misogynistic male recently. There are the lecherous, demeaning sexual predators who have permeated the highest levels of our government. We have learned of atrocities committed by Hollywood elites, Olympic doctors and overly paid TV news anchors. My heart aches for the victims of these crimes, especially for those who reported the harassment and were pushed aside, intimidated and even silenced.

I am often asked about my experience with the men at West Point. In response, I will state it depends on who we are talking about. I certainly experienced harassment and bigotry from upper-class males as well as male classmates. Yet at the same time, I also found some men to be incredibly supportive of women at West Point. Unfortunately, I found more of the former than the latter.

In the few books written by some of the first West Point women, they convey similar experiences. Where we may differ in our recollections is based on our individual experiences at the Academy. For example, the First Regiment had the reputation of more extreme forms of hazing of all Plebes. I know a male classmate who was stripped naked and tied to the clock tower in First Regiment the night of his birthday. And within that regiment were entire companies who pledged to "run out" every female cadet. Existing in a toxic environment like that would color any cadet's experience.

Just as there is disparity among women's experiences with men at West Point, there are differences among the men. There were men who believed that the only reason a woman would attend the Academy was to "find a husband." I would laugh at the absurdity of that statement. We got sweaty and dirty along with the men. The uniforms were never becoming on women. When we donned our gas masks and military gear, we were not appealing in the least. Going through all we had to go through just to gain a husband was ridiculous. However, this belief was pervasive.

There were men who believed that the only reason a woman would attend the Academy was to "find a husband."

Then there were the others who loved to flirt with women cadets but would never date us. Some would deride those men who did date us, implying they were somehow desperate. I believe there were also men who were intimidated by strong women. Others were shy and simply would not ask us out. Many became our brothers. They would be there for us and encourage us, just like any other cadet. This is why I often jest I did not experience a college sorority, I was in a fraternity!

To many men, we were alternatively ugly and fat bitches, authoritarian lesbians, or loose women seeking a husband. At best, we were tough broads or brainy but lacking in leadership ability.

There was a small minority who saw us as equals and colleagues. These men broke with the majority, dated, and even married us. I am pleased to report that there is currently a Long Gray Line of successful West Point graduate marriages. Their legacy even includes having children who went on to attend and graduate from the Academy.

I will be the first to point out that West Point has significantly evolved since I attended. Sexual harassment training is required. There have been investigations with consequences. There are more women admitted and more women in the Officer Corps to act as role models. Although no institution is perfect, USMA continues to evolve and improve in sexual relations maturity.

However, in those early years, I had enough negative experiences that I would be remiss to not mention a few that stood out. One such experience began when I was a New Cadet in Beast Barracks.

Beyond all the cadet knowledge we were required to memorize and "spout off" upon demand, my squad leader composed personalized Plebe poop for each of his charges. I believe the scripts reflected how he viewed us and gave us some levity from our rather oppressive existence. I do not recall what each of my squad mates were required to memorize, whether male or female – but they were all satirical. Mine read:

> "Sir, I am a lean, mean fighting machine. I'm 98 pounds of twisted steel and sex appeal. John Wayne would rather French kiss a rattlesnake than mess with the likes of me!"

Initially, I was embarrassed to have to proclaim that I had sex appeal. In reality, my hair was shorn off, and I wore military-issued glasses and an ill-fitting uniform. I was anything but sexy. Yet, we were required to keep repeating our "poop" until we could shout it with abandon. And soon, several upperclassmen would clamor around us and demand that we repeat our personal description over and over.

In today's thinking, would we categorize my personal pronouncement as sexual harassment? Probably so.

In today's thinking, would we categorize my personal pronouncement as sexual harassment? Probably so.

At the time, however, I was somehow elated. At least my squad leader recognized I was somehow tough and unpredictable like a rattlesnake. When everything "civilian" has been taken away from you, you appreciate being treated as an individual.

During the academic year, we had only ten minutes to get from one class to another. A cadet could not be late, or we were awarded demerits. The academic buildings are quite far apart on our sprawling campus. Once the academic year began, we no longer needed to salute upperclassmen; however, we still needed to acknowledge every upperclassman that we encountered. On more than one afternoon, as I scurried to another class while carrying an armload of books, a throng of male cadets approached. As I came within earshot, I would articulate loud enough to be heard, "Good afternoon, Sir. Good afternoon, Sir."

Many of those male cadets would not acknowledge my effort. On that day and on several others, the upper-class male cadets sarcastically replied, "Good afternoon, bitch." Or, as one of my woman classmates encountered, "Stop staring at me, you ugly bitch!"

At the time, the thought of reporting this form of sexual harassment would have been pointless.

At the time, the thought of reporting this form of sexual harassment would have been pointless. As Beanheads, we accepted that we had little authority. Reporting an upperclassman was rife with consequence. We could be further targeted for intense "questioning." Also, given that we were moving so quickly, we could not even read the tormentor's nametag. Reporting such offenses to our company tactical officer, who typically was a male Army captain, was never encouraged. Most viewed the TAC as someone in authority who awarded punishment and not someone who championed a woman cadet's plight. We simply endured the harassment.

Unfortunately, this was not always limited to upperclassmen. I remember when I was a Senior (or Firstie), our tactical officer had each of the four classes represented in our company complete a culture survey. The results depicted an atmosphere full of dissension, with most women feeling ostracized and devalued. I read the results with remorse. As I was a Senior, I felt compelled to see about making improvements to benefit the experience of the younger women in the unit. Using my chain of command, as we are taught, I went first to see the company commander, who was a classmate. I remember entering his room and explaining my dismay at the survey. I expounded that, as Firsties, we could still make some lasting improvements.

As I spoke, I noticed that he was sitting on his desk with his feet on the seat of his chair. He was wearing a ragged T-shirt and some old tattered shorts. His body language indicated he was disinterested. I continued, ramping up my conviction about the need for change. As I spoke, I noticed he had reached for his groin where there was hole in his shorts. He then reached through the hole. I was not sure if he had a critter in there or was actually enjoying himself; I am not even certain he was aware of what he was doing. It was like I was in the locker room with a jock fondling his manhood. The more he continued, the more disgusted I became. I eventually curtailed my statement and left.

It was this locker room environment and immature leadership that resulted in the dismal cultural findings.

In retrospect, he was probably a typical twenty-two-year-old male, unaware of his ball scraping tendencies. However, I was angry that this was the person who was given the top leadership position in the company. Ironically, it was this locker room environment and immature leadership that resulted in the dismal cultural findings. His apathy destined the company culture to remain unchanged.

In that same company, however, there were some genuinely positive, encouraging male leaders. Historically, all men were required to take both wrestling and boxing at the Academy. There was little thought given to how boxing sparring partners were assigned, rather, the assignments appeared to be random. Depending on the size and skill of your boxing opponent, a male cadet could be pummeled for the entire bout. For some, the barrage of repeated blows to the head made it extremely difficult to "concentrate" in an academic class held later in the day. Others loved the thrill of the ring.

When women arrived on the scene, the DPE decided we should be taught self-defense rather than boxing and wrestling. We learned numerous moves to defend ourselves and inflict injury on an assailant.

After completing the class, the DPE instructors introduced "City Streets," a class where we might practice these skills on our male classmates. However, similar to how boxing sparring partners were picked, our partners were randomly assigned.

I was paired with my company-mate Larry Beisel. Larry was 6'2" and weighed 220 lbs. and was a heavy weight on the West Point wrestling team. As a petite woman, I was expected to throw my sizable aggressor over my shoulder, land him on his back and then kick him with extreme force. As we awaited our turn to be graded, my herculean classmate whispered, "Don't worry, Sara, just yank on my arm and I will do the rest." When our time came, I heaved his arm over my shoulder, and Larry hurled himself up and over. And with a loud "splat," his back contacted the rubber mat. The sound reverberated across the room.

Female cadets in self-defense class 1980-81. Photo courtesy of Lorraine Lesieur, USMA '83

Quickly, I maneuvered his arm, forcing his back up off the ground. I then kicked his back with the front of my shoe so as not to harm but to demonstrate that I could. The DPE instructor observed with satisfaction. Convinced of our theatrics, I was awarded an A+ for the class.

Larry could have allowed me to fail this insurmountable test. He did not. Instead, he found a way to work with and encourage me.

Larry could have allowed me to fail this insurmountable test. He did not. Instead, he found a way to work with and encourage me. We actually had a lot of fun from there on out, trying to make our sparring even more convincing. I got better at the skills, and he taught me ways to leverage his body weight. His leadership was shown in assessing the situation and determining how he might assist. His actions furthered my proficiency in self-defense. The interaction ensured a mutual respect between us.

There were times when situations required the intervention of the officer cadre. As a Cow (or Junior), we had more privileges. I attended an event off post. During the event, a captain offered to drive me back to Highland Falls where I was to meet a friend. Academy regulations prohibited personal automobiles until we were Firsties. So, I innocently accepted his offer. As we drove, he asked if we could stop by his apartment. When we arrived, he asked me to come inside as he wanted to show me something. I was trusting and naive. We entered his apartment and he grabbed his guitar and began to try to serenade me. I became very uncomfortable, and to make matters worse, he was blocking my way out of the room. I simply stated, "Sir, I need to leave." Fortunately, he did not force me to stay.

After mentioning this experience to some of the members of my cross-country team, word reached my regimental tactical officer, a lieutenant colonel. He called me into his office and asked me what had transpired with the captain in question. Apparently other women cadets had shared similar encounters with this man. The regimental tactical officer listened carefully and thanked me for coming forward. The offending captain was quickly confronted and removed from the Academy. In this instance, it appeared that there were officers wanting to ensure that woman cadets felt safe. As a woman, I could report sexual harassment and the offending man was held accountable.

I am sure there were many incidents of sexual harassment that were never reported. The fear of reprisal or a lack of trust in their leadership may have kept more women from coming forward. As I learned with my classmate, (the company commander with the hole in his shorts), some women's words may have fallen on deaf ears.

Lieutenant Gray was a male classmate who had evolved in his support of women at the Academy. He shared with me an experience he had while attending Armor Officer Basic Training. He had a captain as an instructor who informed his students that he was a 1979 Academy graduate, adding that it was the last class with "balls." This was a reference to the last Academy graduating class that was all male. The captain then began a series of women cadet jokes. Lieutenant Gray had recently married a woman graduate. During the class, he became increasingly infuriated as this man began to disparage women cadets by insinuating they were all fat and physically unfit for the Army. After class, LT Gray approached the captain and stated, "My wife is a graduate. She is not fat and could probably do more sit-ups and push-ups than you!" This left the instructor dumbfounded, but he shrugged and simply walked away. Here was a male classmate who championed women cadets!

"My wife is a graduate. She is not fat and could probably do more sit-ups and push-ups than you!"

When I attended the Academy, there were many arrogant West Point men. Those who let you know they were smart, good looking and capable. And then they entered the Army. I soon learned that the "ring knocker" moniker of a West Point officer could be a label of derision, a derogatory statement implying Academy graduates were in an elite club of the self-important. This is a label I hope many of us have worked hard to dispute.

Over the years, I have worked with many men, some of whom have been West Pointers. One 1972 graduate became a mentor, grooming me to become a management consultant. Bob Bryant saw my potential before I recognized it within myself, and I am indebted to him to this day. Initially, Bob gave me a client to coach and had me create some leadership training programs, and then I learned to develop my own clients. This eventually led me to a large contract that lasted for five years.

Soon after meeting Bob, and with his encouragement, I began a networking organization for Academy graduates in the local Richmond, Virginia area. The group was comprised of West Point, Navy and Air Force graduates. Our group became a resource and referral organization for its members. I formed mutually supportive relationships with many of these men and women. Some were quite generous with their time and expertise as I grew my company. I assisted several of them with contacts while searching for jobs and resources to expand their businesses; in turn, many acted as viable resources for me.

Another experience with a West Pointer proved disappointing. I formed a relationship with a company led by a couple of West Point men to earn a large contract. In doing so, I became a subcontractor and they became the prime. I quickly learned that the prime would default to their positional authority whenever a disagreement occurred. In one heated exchange, the prime who had been a battalion commander in the Army reverted to an authoritarian voice and demanded to know what I had earned in my previous roles. This question violated a number of employee relations laws. Eventually, our disparate views led to the ending of our professional association. However, I had learned long before to reject unacceptable behavior even from my Academy brethren.

I have come to conclude there will always
be men and there will always be "lesser" men.

Over the years, I have connected with some of the West Point men from my Academy years. Most have grown and matured and have become much more supportive advocates of women. There will always be a few who revert to intimidation and condescension. But, I have come to conclude there will always be men and there will always be "lesser" men.

CULTIVATING CHARACTER

I have spent years teaching and writing on diversity and inclusion. I lived a real-world and very intense diversity experience at West Point. I find that engaging men, especially white males, in the conversation is the only way forward.

Creating an environment where all individuals feel safe to bring the uncomfortable topics to the forefront requires that we do not limit participation. Diversity or employee resource groups that are not open to everyone work against dialogue and prevent hearing opposing viewpoints. Because gender, race and sexual orientation topics are often quite challenging, having an independent, trained facilitator is an excellent practice.

From the individual perspective, one can always choose how to respond to a gender-biased or derogatory comment. At times you may overreact. At other times, it's necessary to shut down or ignore an inflammatory statement or an inappropriate remark or proposition. Learn how to respond assertively and appropriately the next time.

- Say "ouch!" when a statement offends you. Often an offender will be forced to think about what was said and may be less likely to repeat it simply because one person challenged him or her.

- Another approach is to restate the offensive statement or suggestion. "So, you want me to …?" The offender will then be forced to explain their statement, which often will force them to re-think the original words.

- Always, always, always report inappropriate behavior. If nothing changes, seek legal advice.

There were good men and flawed men at West Point. I hope most of the chauvinistic men I endured have matured. The Academy has made many strides to reduce sexual harassment, and continuing these efforts is required.

In the workplace, respectful dialogue that engages all parties on gender issues is the best means forward.

ON SHATTERING PRECONCEPTIONS

Beyond the challenges of working with men, women will continually be forced to shatter preconceptions.

In truth, I came from a non-conventional home. Unlike many of my peers, both my parents worked and had professions. My mom, Estelle or "Tel" as she preferred to be called, defied many conventions of her time. She pursued physical education and eventually earned a PhD. When she landed a professorship in Boston, my dad supported the move, seeking his own employment after the relocation. My parents divided household chores, with my mother doing most of the household cleaning and my

Estelle (Tel) Miller Fotsch, the author's mother

dad doing the cooking. This latter fact might be the reason I was such a skinny kid.

My mother was also the consummate athlete. She was skilled in such divergent sports as tennis, water ballet, sailing and skiing. She would tell us that if she hadn't married and had kids, she would have been a ski bum in Colorado.

As a physical education specialist, Mom was a big supporter of Title IX, ensuring that both male and female students were granted equal opportunities to play sports, achieve athletic scholarships and have equitable facilities. She was also an excellent typist. She typed not only her PhD dissertation but also my dad's, all while raising my three brothers and me.

As evidence of the competitiveness of my family, when my parents were in the their sixties, my brother Sam called and excitedly related, "I finally beat Mom and Dad in tennis!"

All her kids were involved in several sports, and as a family we played tennis together. Many a Sunday afternoon, we would head to the public tennis courts and play. Individually we might be paired against Mom and Dad. They played to win; my mom might play the net and my dad would put back spin on his return shots. The combination assured they would win game, set, and match more often than not. As evidence of the competitiveness of my family, when my parents were in the their sixties, my brother Sam called and excitedly related, "I finally beat mom and dad in tennis!"

Having three brothers, I competed right along with them in all types of sports. I am certain that playing many hours of basketball, baseball and tether ball with my brothers is where my competitive spirit was born.

My mom could have a quick temper. When she was driving, if another motorist cut her off, she would beep her horn and yell out the window, "You fat head!" I assume this is where I learned candor.

When I might bemoan that my clothes were not as stylish as my friends' outfits, she would admonish me, "Sara, be your own style maker." I did not know it at the time, but my parents, and especially my mom, breaking with convention served me well as I entered the male dominated bastion of West Point.

My mom breaking with convention served me well as I entered the male dominated bastion of West Point.

There were moments at the Academy when I was acutely aware that I was atypical, often running up against conventions and definitions of what a woman should and should not be.

One of the first memories of this phenomena was with my second detail Beast Squad leader.

At the halfway point in Beast, the new leaders (typically Seniors, or "Firsties"), would take over as the cadre. Those squad leaders were referred to as a "second detail" squad leader.

Cadet "Tall & Handsome" was a demanding leader. I remember him marching next to me on a road march and asking me to again recite some Plebe "poop" that I had not yet done successfully. This was a forced hike where we carried a heavy rucksack and our weapons for miles over hill and dale. The march was physically taxing. However, the rhythmic

Schofield's Definition of Discipline

The discipline which makes the soldiers of a free country reliable in battle is not to be gained by harsh or tyrannical treatment. On the contrary, such treatment is far more likely to destroy than to make an army. It is possible to impart instruction and to give commands in such a manner and such a tone of voice to inspire in the soldier no feeling but an intense desire to obey, while the opposite manner and tone of voice cannot fail to excite strong resentment and a desire to disobey. The one mode or the other of dealing with subordinates springs from a corresponding spirit in the breast of the commander. He who feels the respect which is due to others cannot fail to inspire in them regard for himself, while he who feels, and hence manifests, disrespect toward others, especially his inferiors, cannot fail to inspire hatred against himself.

Major General John M. Schofield
Address to the Corps of Cadets
August 11, 1879

cadence of the march helped me remember Schofield's Definition of Discipline.[1]

I recall that when I eventually completed the recitation accurately, Cadet "T&H" turned toward me and simply stated, "You are one tough cadet." Since compliments were extremely rare, I was elated. Still, I also heard a slight lift in his voice, as though it was a bit of question. It made me

---

1   Graham, D. R. (n.d.). Bugle Notes: Learn This! Retrieved March 17, 2018, from http://www.west-point.org/academy/malo-wa/inspirations/buglenotes. html.

understand that because I was one of the women to complete the march, I was a conundrum to him, a bit of a puzzle.

I also remember him advising me to not wear the military issued glasses after the academic year began. We referred to these '50s-style black square glasses as "birth control" glasses. As we were not permitted to wear our civilian glasses, I thought, "Great, I look like a dork. He just confirmed this fact, and I can do nothing about it!"

When you are seventeen-year-old woman, you don't want to hear you are tough and look like a dork. So, I decided to just be happy that I had at least earned the title of being tough.

Cadet T&H could also be encouraging and even surprised our squad by breaking with traditional protocols and allowing us some fun. For fun, my squad leader transformed his room with a disco ball and turned up the John Travolta music on his stereo. He marched our squad in and ordered us to dance for about twenty minutes. My squad mate, Joe, despite my birth control glasses, picked me up by my arm pits and swung me between his legs and over his head. This cataclysmic respite was short-lived, but one of my fondest memories of Beast.

Later in my Plebe year, after we earned a weekend away from West Point, a group of friends and I decided to go to a classmate's home on Long Island. As cadets cannot have a car until our Senior year, we took a train into New York City. We arrived in Grand Central Station adorned in our long gray overcoats and wearing our formal dress gray uniforms and hats. As we waited for our connecting train to Long Island, a woman came up to us and asked us, "Are you in the Salvation Army?"

Embarrassed, we told the young woman that we were West Point cadets. I felt even then that we were not to be understood easily by our civilian counterparts. At Christmas, I was asked to speak at my church along with other recent graduates. Again, I wore my dress gray uniform. A former cheerleader from my high school came up to me and commented, "I really like your outfit." I simply said, "It is called a uniform." Inside I was seething and resented not being in a cute outfit like the one she was wearing.

Later I was working out with a male classmate, preparing for another Army Physical Fitness Test (APFT), which included two minutes of push-ups, two minutes of sit-ups and a two-mile run wearing combat boots. The topic of the different physical standards for men and women at the Academy came up and we began debating.

Cadet Gray asked how many push-ups I had done at the last APFT. I, in turn, asked him his total. I had scored more push-ups than Gray. Taking this in, "G" responded, "Yeah, but the DPE instructors are not as hard on you women. They don't make you 'break the plane' like they do with the guys."

Break the Plane

Break the plane refers to lowering the body from the front leaning rest position so that the upper arms are more than parallel to the floor (i.e., "breaking the plane"). However, if a cadet fails to break the plane, the push-up would not be counted. It was typical for DPE instructors to not count several of our attempted push-ups. One could hear them counting, "1, 2, No, No, 3, 4, No, etc." The "Nos" would be demoralizing and tiring. The push-up completed in this manner required a great deal of upper arm strength and many women struggled with this challenge.

To view how it should be done go to:

https://www.youtube.com/watch?time_continue=11&v=Lu4JpdWTEhQ

I replied rather indignantly, "Yes, they do. I have to break the plane, just like you do."

So, we decided to test our theories and challenged one another to a contest. As I began knocking out push-ups along with him, he began to see that I was surpassing his count. After the two-minute effort, he sat back on his heels, sighed, and reluctantly admitted, "I guess you *can* do those push-ups." Of course, I thoroughly enjoyed that moment. Fortunately, he was an atypical male cadet. He was not threatened by my physical aptitude. In fact, we continued to debate and compete regularly in a variety of subjects!

"I guess you can do those push-ups."

Despite my progress at the Academy, I found I was still an enigma with members of my own family. My Aunt Aileen from Arkansas came to my West Point graduation. During graduation week, the First Classmen could host four of their family members in the Mess Hall. This meant my family occupied half of the table and my male counterpart and his family the other half.

Having arrived early, I sat at the head of the table and my male class-mate sat at the Fourth Class end of the table. He performed the Plebe duties such as filling glasses and slicing the dessert. Near the end of the meal my aunt observed him slicing the dessert. She turned toward me and inquired, "Sara, honey, what is your classmate doing?" I assumed she wondered why the Mess Hall staff had not sliced the cake for us. I explained how he was simply doing the "duties" of a Fourth Classman. She shook her head in disagreement and stated, "Honey, that's woman's work!" I was initially speechless and then felt incensed by her statement. I thought to myself, "What does she think I have been doing for four years?"

My three brothers each struggled with defining their only sister. To this day, they refer to me as "Sara, the West Pointer, Dad's favorite son." I know that they are proud of me; but, still, this is the kind of left-handed compliment I heard repeatedly. You just get used to it.

As a young lieutenant in Germany, our Forward Support Battalion had a German Army partnership unit we spent some time observing, comparing and contrasting our different approaches to military leadership and organization. Early one morning, several of our company leaders met at the German unit's headquarters and set out for a brisk run. The German Army does not have women, and as one of only two women running the event, I completed the run somewhere in the middle of the pack. A young German lieutenant approached me after the run and asked, "Fraulein, did you run from the beginning?"

His question implied he had a hard time believing a woman could run three miles. By this time, I was used to breaking preconceptions and simply nodded my head and said, "Yes!" But, I was thinking, "Yes, and then some - I began my race over four years ago at West Point."

After forty years of having women at West Point,
new conventions are emerging for women.

Now after forty years of having women at West Point, new conventions are emerging for women. Women are now able to enter the combat arms branches of the Army. They can be infantry and armor officers. Women have even earned a coveted Army Ranger tab. The redefinition continues.

A recent experience details just how far we have come. I had several West Point women over for dinner at my home. I was the "old grad" of the group, as most of these women had graduated after 2000. I was a bit in awe of my guests, all of whom had seen combat, and all were mothers. All but one was still on active duty. We enjoyed some wine with dinner and then the stories began.

Carolyn Furdek, of the Class of 2000, shared a story about her young son, who has only known his mother to have been in the Army and his dad as a civilian. At Christmas, her son received a gift of a GI Joe doll with a parachute. It was a toy he could drop from balconies or throw in the air and watch it float down. He opened the present, and as he held his new toy, he squealed in delight and said, "Mom, I love it. What should I name *her*?"

This led another woman, Brittany Meeks Simmons, USMA Class of 2002, to tell us about a similar experience. Her young son only knew his mom to be in the Army and his dad to be in the Air Force. One day he said to his mom, "So women are in the Army and men are in the Air Force?" From the mouth of babes …

Quite possibly we have come full circle and one day we will need to explain that yes, men can go into the Army and women have been in the Air Force for a long time.

When you find yourself facing or exceeding limitations that others place on you, then by all means, be prepared to have your accomplishments questioned or even diminished. At other times, you will need to accept that well-meaning people are simply trying to understand a phenomenon outside their own experience.

Some of my male classmates to this day still believe that most women "got over" or had it easier when it came to the physical demands of the

Academy. For some of them, a woman excelling somehow took away from their masculinity or identity.

In my years of consulting and coaching, I have seen this phenomenon repeatedly. Sometimes, it is when a younger employee excels beyond what their peers or leaders anticipated.

I worked with a talented young woman who earned an MBA from a prestigious university and was quickly moved from individual contributor to manager to director in less than five years. This was an unheard-of trajectory in this company. Most of the current directors took ten to fifteen years to earn a similar position.

In the company's pursuit of having more women in leadership roles, "Sandy" was identified as a high potential and given a more demanding role. However, her aggressive promotions actually hurt her overall success. Before she had mastered managing others, she was put in the position of managing those who manage others, a much more complex leadership role.

In the end, leadership failed to recognize that they had not adequately prepared her for this role. When she did not perform as well as her director peers, they simply let her go.

Only one woman executive spoke up for her: "We are doing something wrong with our talent management if we think we will be successful when we eliminate an Ivy league MBA and we don't examine how we arrived at this decision."

In the end, this woman went on to redefine her career path and it never led her back to the corporate arena. However, the company lost another woman who may have made the ranks had her advancement been planned similarly to that of her peers.

As some of the first women graduates of West Point, we were working against institutional and societal conventions, even while defining who we wanted to become. We ran up against harassment and prejudice unique to our circumstance. For me, it created some defensiveness, yet it also taught me compassion for others in similar circumstances.

CULTIVATING CHARACTER

Do you find yourself breaking with a traditional role? Are people telling you that you are too young and inexperienced or that you are too old and over qualified? You may have a disability that others believe should limit you. Yet, you are convinced it should not hold you back.

Are family members and close friends questioning your career decisions?

How do you contend when you work hard to shatter perceptions and struggle to have your abilities accepted? Here are some suggestions:

• Run your race.

Many cadets, like all human beings, put off training for the annual Army Physical Fitness Test. Like clockwork, as the date of the APFT approached, more cadets were suddenly out running. As a runner, I ran nearly every day, so it was natural for me to run five or six miles at a clip. I would start out running and I would slowly catch up to a male cadet. And when I passed him, he would notice that it was a woman and then start running faster just to pass me. His actions were as if to say, "I can never have a female cadet pass me in a run." I would just keep running. In time, I'd catch up and pass him again. He might try to accelerate once more. Rarely would a guy try to overcome my momentum more than twice. As I passed him, through my peripheral vision, I would see when the cadet was done challenging me. He would not really acknowledge me; his head would bend down, his shoulders might sag, he would just slow down and just let me continue on.

In the same way, when you really know what you can do, you can just keep doing it. In time, even your detractors will be resigned to the fact that you are successful. They may do so grudgingly, but in the end, does it really matter?

- Find a role model who overcame tremendous odds. Meet that person. Read their biography or watch a movie about their achievements. Then, ask yourself what you can apply from their story to your situation. Heroes like Phyllis and my mother, who I mention in this book, are women who continue to inspire me to not settle with convention.

- Learn to accept that some people will never understand your drive or ambition. For some, it threatens their own mental constructs. If insults come, ignore them and reaffirm your direction through positive, affirming people. To learn more about creating your "tribe," see the Encouragement chapter.

- Be sure you assess your own biases from time to time. Refrain from setting limits on others. As I mention in the Humility chapter, the best leaders recognize that people are gifted in many ways and they learn to harness the talents of their entire team.

Be patient and continue to run your race. In time, the naysayers will be a distant memory.

THRIVING DESPITE
THE SYSTEM

Up to this point, my intent has been to describe several of the principles of the West Point ethos. I have outlined the challenges of women cadets and those who were found to be under their rule. I provided a snapshot of the arrogant, chauvinistic male cadets and contrasted the majority with the few progressive Academy men. I also portrayed the preconceived notions of our peers and family. All this was written to provide a backdrop for this chapter: to describe how one can thrive in a barrier-filled environment.

Like the majority of those who attend the Academy, I chose West Point as a means to obtain a college education. However, even the first time I visited the Academy, I sensed there was much more to West Point than a challenging academic program. There was an honor code and that appealed to my moral sense. The cross-country coach really wanted me to run for him, which excited me. Despite all that, I could feel that there was already a deep resentment toward women, and I knew it would never be easy.

In my limited free time as a cadet, I recall walking down a hallway in one of the academic buildings and noticed several portraits of famous graduates. I viewed the likes of Sylvanus Thayer, Robert E. Lee, Ulysses S. Grant, Thomas J. "Stonewall" Jackson. There were no pictures of women. I remember thinking, "Why do I feel so out of place here?"

I remember thinking, "Why do I feel so out of place here?"

With attrition rates averaging 50% for the first several classes of women, there were very few women in the upper classes to use as leadership models. And not all upper-class women were interested in being our mentors. In fact, as I mentioned previously, a common complaint in my class was that many senior women made it *more* difficult on us. The fear of reprisal from male cadets for helping another woman cadet might have prevented their assistance. The Class of 1983 was a sandwiched class, where we were still a novelty dealing with resentment from male cadets while also dealing with, at times, an aloof, disinterested and caustic women cadre. All these factors contributed to the "Proud to Be" Class of 1983 having the all-time lowest women graduation rate (48%) in West Point history. All I knew, as I peered at those famous portraits, was that I felt incredibly isolated and alone.

I will be the first to say that many men also faced difficulty at the Academy. You only need to look at the history of Academy.

Nearly 100 years prior to women arriving at West Point, Henry O. Flipper, an African American, dared enter the Academy. He endured four years of shunning from the entire Corps. Yet Henry bore such flagrant abuse with dignity. He persevered, applied himself to his studies, and in 1877 became the first African American to graduate. He later wrote, "To stoop to retaliation is not compatible with true dignity, nor is vindictiveness manly."[1]

The prejudice continued into Flipper's Army career. As an officer, he was brought up on trumped up charges and discharged from the Army.

---

1   Cowley, R., & Guinzburg, T. (2002). *West Point: Two Centuries of Honor and Tradition.* p. 119.

Despite all these forces against him, he chose to press on. He wrote and edited papers on the Southwest Territory. He was hired by William F. Buckley Sr. to consult on a Venezuelan oil venture. Later, he became the assistant to the Secretary of the Interior. Eventually he published his autobiography, *A Colored Cadet at West Point*. Henry chose to thrive despite the biased reality of the system in which he lived. Posthumously, his name was finally cleared of all charges in 1999.[2]

Henry chose to thrive despite the biased reality of the system in which he lived.

Susan Kellett-Forsyth, a graduate of the first class of women, provided the following example of the vitriol the first women cadets endured from their male counterparts:

One night at about midnight, Sue was roused from her sleep by upper-class male cadets pounding on her door. They demanded that she get up, get dressed and "report" to a specific room. She recalls that in that room there was a long table with a gray West Point blanket on it. There may have even been candles. She had to stand at attention against the wall lockers and receive a lot of passionate explanations about why she should not be at West Point. Fortunately, she did not let this traumatic incident deter her; however, she was the only female cadet in her graduating class to endure an entire Plebe year in Company A-1.

There was evidence of an undercurrent within the faculty of making things more difficult for women. Because there were so few of us, we tended to stand out from our male classmates. I remember several professors seeming to call on me more frequently than my male cadet classmates. I was truly afraid of looking incompetent. Being called upon

---

2   Ibid.

seemed to happen more often than not. As a result, I shut down and did not participate in many classroom discussions; in the end, that likely hurt my academic performance.

I had one professor (or "P" in cadet parlance) who gave a library assignment and posted it in the back of the room on a bulletin board (yes, this was eons before the likes of online Blackboard). We were to write down our assignment, then go to the library and complete it before the next class. I hurried to the back of the room, with men clamoring around the print out and trying to read it. Then I exited quickly to get to my next class on time (demerits were awarded for being tardy).

In my rush, I had written down the incorrect assignment. I turned it in as requested. When I got my grade back, it was an "F," and added to the grade was a demerit slip for "failure to follow orders." I ended up confined to my room for a weekend, all because I did not write down the assignment correctly.

To this day, I am not sure if this particular "P" did this with others in the class. I have shared the story with other classmates and they have never had a similar experience. The point is that there were enough instances like this, that I felt targeted, made an example of and diminished.

In most universities, feedback can be given on professors, those who consistently receive negative criticism from their students are not allowed to continue to teach. In addition, most colleges allow students the ability to change instructors. In my time at the Academy, a cadet had very little control over his/her schedule. I once set up a meeting with the head of a department requesting that I be moved to another professor's class since I had not been successful with that "P" in the past. I suggested that if I had a different instructor, I may do better. The lieutenant colonel simply looked at me and stated in no uncertain terms, "Cadet, we don't make changes to accommodate just one cadet." No discussion–no option.

On the flip side, there were dedicated professors such a French instructor who made ample time to help those of us who struggled. There was also a math "P" who provided tutoring long after normal work hours. However, it was my experience that these approachable, encouraging professors were few and far between. More often, I would seek out class-mates to assist me from time to time. Beyond that, I just buckled down and studied for hours on end.

I would seek out classmates to assist me from time to time.

To make matters worse, the core curriculum was completely based upon achieving a general engineering degree. We were required to take courses like mechanical engineering, thermodynamics and electrical engineering. I did not enjoy these courses. We could not major in a different subject area and were only allowed a few courses in a concentration more to our liking. I chose behavioral science and leadership and was enthralled with these classes. My only regret was that I could not spend as much time studying the classes I loved, since the other less enjoyable classes required a great deal more of my time.

As I have mentioned, even with burning the late-night oil, I was only an average student. I found solace in competing in cross country and track and eventually with the Army Marathon Team. In the end, I was known more for my brawn than my brain. I was eventually named the top woman athlete in my class – meaning my cumulative scores on my physical education courses resulted in me earning this distinction.

My male classmate Gray also struggled academically at West Point. He loved math and chose it for his concentration. A glutton for punishment, he took increasingly more difficult math courses beyond those required for the rest of us mere mortals. I remember him spending four or more hours working on a single math problem.

We both made the most out of a challenging academic scenario. The determination we needed to graduate served us well in our careers. Neither of us were afraid of hard work. Gray earned distinction in an engineering career, and I was able to develop a highly profitable consulting business. When people would ask where I went to school, they would never follow up and inquire about my grade point average. They assumed I was reasonably intelligent simply because of the association with West Point. Now I take pride in the fact that I was in the third of the class that made the upper two possible!

Later, when I was an army captain attending my advance course, the ordnance school commander would regale us with stories of his various

Promotion to Captain with other West Point
classmates at Ordnance Advance Course, 1987

assignments. He had inherited an Army recruiting command that was flailing. His recruiters would complain that their competitors, Navy and Air Force recruiters, had many more incentives to offer than they could offer the young men and women who entered his command's offices. Unlike a sales organization in the civilian world, the commander could not incentivize his staff with money or promotions. Instead, he adopted a policy called, "Made Mission–Go Fishing!" As the slogan implies, once his recruiters met their targets, he gave them complete freedom in how to use their time. He released them from a strict work schedule and rewarded his performers with time off. With this and other changes, his command began to excel. He found a way to thrive despite the system.

A few years ago, I was hired to assist a large company with change management. Initially, the company was overrun with numerous external consultants, many of whom did not want to play with the new internal consultant (me). To further complicate things, the company had been through multiple reorganizations. My role was reassigned to five different bosses in three different parts of the company all within two years. My last boss, despite the constant chaos and rate of change in the company, saw no value in change management. I felt stuck, undervalued and frustrated. Given this scenario, how did I choose to thrive?

Earlier on, I heard that the CEO had launched a Lean Six Sigma process improvement program. So, I jumped in and assisted in developing the program using an employee engagement process. Although Lean Six Sigma was not my area of expertise, I showed up and contributed, and in little over a year, I led the program.

The shuffling didn't stop. And one of my bosses asked to me to serve on the Veteran Steering Committee and work with some other very dedicated veterans and develop the company's first Veteran Employee Resource Group. The group became so successful that it became the model for all the other ERGs in the company. It was during this initiative that a colleague told me about a state program offering company certification as a veteran-prepared and educated company. I also learned that the state program was coming up for bid. When I read the requirements of the request for proposal, I knew I had the skill set to do the work.

I partnered with a larger firm, and we won the work that lasted for multiple years. Through persistence, I had developed an exit plan to leave a system where I was stymied. I focused on what I could control and did those things extremely well. This led to new opportunities. In the end, I eventually developed a much more rewarding and profitable career path leading to my own consulting practice.

Through persistence, I had developed an exit plan to leave a system where I was stymied.

A successful executive friend shared about a time when he was ready to throw in the towel. "Mark" was vying for a promotion to manager along with two co-workers, "Ben" and "Jane." Their superior promoted Jane over Mark and Ben. Both men were convinced they were a better choice for the role, and they were deeply disappointed. However, how each of them handled their disappointment was quite different and provides an object lesson.

Ben was vocal about his disappointment with the decision, and he even began to work against Jane and the "system" that had promoted her. Mark did not complain publicly. He realized that his superiors would take note of a negative outburst. Instead, he chose to work with Jane and continue to develop his own leadership skills. A few years later, he was also promoted to manager. A few more years later, Mark was selected over Jane for a division manager position. Ben's career, on the other hand, stagnated and he remained a supervisor until he retired.

We will all face demoralizing and difficult scenarios in our careers that can make us feel stuck. The key is to recognize when we feel trapped and to develop methods to work through these obstacles. We must work against allowing the "system" to dictate our attitudes and behaviors, and we need to apply precepts we have learned thus far to forge ahead.

We must work against allowing the "system" to dictate our attitudes and behaviors.

CULTIVATING CHARACTER

What do you do when you sense that you are paddling upstream and no relief is in sight? Worse yet, what if you feel *completely* stuck?

Below are some insights I have gleaned:

- Pray the *Serenity Prayer*.

- Review the West Point principles described in this book, checking to see if you are applying each tenet to your current situation.

> **The Serenity Prayer**
>
> God, grant me the serenity to accept the things I cannot change, the courage to change the things I can, and the wisdom to know the difference.
>
> **Reinhold Niebuhr**

- Seek ways to demonstrate that you are a team player. Look for ways that you can contribute *within* the system. For example, if the department schedule is constantly changing and creating havoc, ask to take over that role.

- Volunteer for something unrelated to your current role. Having a mental break from the dysfunction can be healthy.

- Play the long game because most situations are transient. Adopt the mentality that this too shall pass.

- Relieve frustration and stress through regular exercise, deep breathing and healthy eating habits. Avoid using alcohol or drugs for relief, as we know that these types of short-term fixes can add up to devastating outcomes.

- Reach out to your positive connection of peers for perspective, suggestions and knowledge of other opportunities.
- Love on your family, and if you have them, love on your pet(s).

In my experience, it is not easy to thrive within a system that is intent on holding you back. I have learned that with practice, it does get easier. By consistently applying the principles contained in this book, in time, you can be successful in navigating those forces too.

PART IV
Digging Deeper

The last three chapters of this book describe leadership tenets that demand greater maturity and commitment. Through very personal and painful experiences, I hope to teach you about love and leading and moving beyond death and loss, and finally I will challenge you to leave your own leadership legacy.

ON LOVE
AND LEADING

There are many types of love. The Greeks, however, granted us a deeper understanding of love by describing it using various forms:

- Philia – Affectionate, warm, and tender platonic love. A brotherly form of love.

- Eros – Passionate and intense love that arouses a romantic feeling, such as one has for a mate.

- Agape – Unconditional and sacrificial love, the kind that bears all things, believes all things.

I can say without a doubt that I experienced all the forms of love at the Academy.

PHILIA

As I have shared through the past pages, the shared misery of Beast Barracks and Plebe year created ties with classmates that have lasted a lifetime. The endless drills and parades created a desire for precision in our military processions and a sense of pride for our cadet company. The honor of competing for the Army team as a varsity athlete imbued a love for my teammates and for competition. The weekend football games, culminating in the annual Army-Navy contest, developed a love for the unity of the Corps. The obstacles that all cadets faced, whether physical, academic, emotional and even spiritual—all these completely unique experiences developed a unity and an abiding affection for one another.

I had many Academy "brothers and sisters" who I found I could rely upon in my years at West Point, yet I was also a young woman who desired romantic love.

I had many Academy "brothers and sisters" who I found I could rely upon in my years at West Point.

EROS

With a ratio of ten men to every woman at West Point, one would expect there would be several opportunities for me to date another cadet. However, at the time, as I have detailed earlier, there was a great deal of animosity toward female cadets. We were ugly, fat undesirables. Men who took the risk of asking us out might be ostracized by their male peers. Others viewed us more as sisters. I ran into several men who loved to flirt with women cadets, yet their advances never culminated in anything more. Often, most of us were simply too busy to have much time for dating.

Also, there were long-held traditions of the Corps that made romance difficult. Cadet Hops, for example, were dances that the institution would host for each class. The Academy, for many years, would bus in women from the surrounding colleges for these socials. This practice continued even after women cadets were admitted. In the early years of women at the Academy, we were required to wear our hair short. It could not touch below the collar of our shirt. To make matters worse, there was no resident hair stylist available, just barbers. I remember members of my track team joking that you could always tell a female cadet because our hair styles were identical. Our mop looked like a bowl had been placed over our heads and the barber simply cut around the edges.

We wore uniforms to a Hop — often a white shirt with gray pants. And if we dared, we could show our legs with a gray skirt that modestly hung just below the knees. Added to that ensemble were our shiny black pumps. Those high heels not only hurt but reminded me of the shoes my mom wore in the 1960s!

Our women peers from the colleges such as Ladycliff had no such restrictions on their attire. Often, they would enter wearing sexy, form-fitting dresses. It was easy to understand why a West Point woman was left to observe the dance from the sidelines.

The few times I was asked to dance, it led to an awkward conversation. The guy might say, "So, I am a West Point cadet." And I would say, "Yes, so am I." He might say, "Well, I'm studying engineering." As the entire Academy curriculum was engineering-based, I would also say, "Yes, so am I." And then the cadet would realize he was not impressing me and would move on.

Even with all these obstacles, it was inevitable that cadets, like most of their peers, would date and show affection.

Cadets were also to avoid "public displays of affection" (PDA) or we would receive demerits. This meant no public hugging, kissing, etc. Also, whenever a member of the opposite sex was in the room, cadet room doors had to be propped open with a trash can. However, even with all these obstacles, it was inevitable that cadets, like most of their peers, would date and show affection. We simply learned where we could meet and have some privacy.

Plebes were also prohibited from dating anyone except for a classmate. But, once an upperclassman, one could date another upperclassman, but not those in the chain of command. An upperclassman was also prohibited from dating a member of the Fourth Class. However, the heart wants what the heart desires, and many a liaison occurred around these restrictions.

For me, I seemed unable to obtain or hold a boyfriend. When I was a Plebe, I dated a classmate. He wanted only one thing, and so that was short lived. As a Yearling, I was pursued by a Firstie. "John" was a very kind, loving man, but I found that over time he truly wanted a "little wife." I was never going to fit the bill. After he graduated, we came to an impasse as a couple, unable to find common ground on what we wanted as a couple and as individuals. While he was in Ranger School, he wrote me a letter ending the relationship for good. This is ironic because some guys complained about receiving "Dear John" letters from their girl-friends back home. I got the reverse experience, a "Dear Jane" letter from someone who was in the thick of one of the most mentally and physically taxing programs the Army can offer.

After the breakup, I was hurt by the rejection. I was livid about how the message was delivered. Yet I was also relieved. I knew I could not be someone who I was not, and ending things was the right thing to do.

West Point is not kind to those who are suffering from a breakup. The pace of the academic requirements never lets up. Suddenly, I was single and alone again. It was one of my darker chapters at that Gray Institution.

Unable to hold onto to a man, I decided to take my mother's advice, date around and not marry until I was thirty. As I approached my final year, I decided to just have fun.

During the summer before my Firstie year, I headed to Camp Buckner to lead a platoon of Yearlings. As Seniors, we had more freedoms including the ability to have a car, and thereby, the means to leave post.

One weekend that summer, a squad mate from my Cadet Basic Training asked me to go dancing off-post. I was to meet him at the Firstie Club and we'd head out. I put on my civilian clothes. As I approached the club, a male classmate sitting outside the club got my attention.

In a propositional way, he said, "Well, hello there!" to which I replied, "Hello?" I was used to faux flirting and so I kept walking. He then asked, "Don't you know who I am?" I stopped and looked at this cocky man and replied, "No, should I?" And then without hesitating, he exclaimed, "Well, everybody knows Eddie Gaba." At this point I was certain I had just met the most audacious of male cadets.

He then began to ask me if I knew the person he was sitting with, and I said, "Yes, he looks familiar." Eddie then held out a dollar bill and asked, "Why don't you go get us a couple of sodas?" To which I replied, "Why don't you go get them yourself?" He replied, "Don't you know, you cannot go into the Firstie Club in a PT uniform?" I looked at him, thought for a moment, and then went in, bought the sodas and gave them to him. Little did I know that this brash young man would become my husband.

Eddie was fond of interrupting me at this point in the story by adding, "It is a very good thing that Sara got those sodas for me or things may have turned out differently."

At that moment, however, I never really thought much about that interaction. I assumed it was just one of the many teases from a male classmate that never materialized into anything more. I went out, danced with my friend and had a good time. Yet, Eddie would find ways to run in to me that summer. Once, I was coming out of the post office and he saw me and said, "I followed you in your cute blue Mustang. At a stop light, I pulled up beside you and looked over and then I realized it wasn't you."

To which I thought, "He's weird and funny, but he is probably not serious."

It was not until we returned to academic classes that fall that Eddie asked me out. I was leaving Thayer Hall as he was going in. He stopped me. A pool of male cadets seemed to stop at the same time and surround us. And then he said, "Sara, will you go out with me?" To ask me out in front of several male cadets took chutzpah to say the least. And I replied, "Yes," and hurried to my next class, the whole time thinking, "what was his name, again?"

After going out with this unambiguous, cocky and handsome man, he totally messed up my plan to wait until I turned thirty to marry. We enjoyed a whirlwind romance and I knew he was the one for me. As graduation approached, we found ourselves deeper in love and began talking about getting married. The Army allowed for a joint domicile or the ability to be assigned within fifty miles of one another provided we were married. After discussing the matter at length, we decided to get married shortly after graduation rather than risk not being assigned together.

At spring break, Eddie, in his deliberate manner, announced to my parents, "Sara and I are going to get married right after graduation rather than wait." My dad seemed unable to accept the finality of this declaration. He dropped his fork, his chin trembled, and he blurted out, "I am not going to go broke for a wedding like Uncle George. And I don't think it will last more than six months."

Obviously, this was not what we had hoped would be my father's response. I believe that Dad was simply reacting to Eddie's abrupt announcement, and he did not like the idea of his little girl leaving. His comment about my uncle was ridiculous. My uncle had seven daughters and understandably had significant financial obligations to pay for their

Sara and Eddie's wedding, May 26, 1983

weddings. Whereas my dad had but one daughter. In time, Dad let go of his curmudgeonly ways. He came to respect Eddie and even compliment his work ethic. We did, however, enjoy reminding Dad, year after year, that Eddie and I had surpassed that six-month mark many times.

I wed James Edwin (Eddie) Gaba, Jr. at the West Point Protestant Chapel on May 26, 1983, the day after our graduation. All the groomsmen donned their officer Army blue uniforms. I wore my mother's wedding dress. My bridesmaids, all but two of whom were women classmates, wore feminine flowing gowns. In military fashion as a newly wedded couple, we passed under the arch of the sabers of several officers. Then, as is the custom when the bride exits, the last officer swatted me on the derriere for good luck.

As you may or may not have surmised, the Cadet Gray and Lieutenant G mentioned throughout this book was the one and same Eddie Gaba.

Neither Eddie nor I were stellar academicians at West Point, and we did not earn high positions in the cadet chain of command. Considering his average performance, Eddie would sometimes bemoan the fact that he was not "known for" anything at West Point. I would quickly remind him that he was known for getting the gal who was known for her brawn rather than her brain. Which, of course, made him chuckle and realize we were both incredibly blessed to have found one another.

Even at a place called West Point, I had discovered romantic and abiding love.

Even at a place called West Point, where I had experienced ridicule, failure, discouragement and loneliness, I had also discovered romantic and abiding love.

AGAPE

The West Point experience created a deep love for our country and for service to others. We were reminded that we needed to care for one another, the soldiers we would lead, and the officers we would serve. In short, we were to take love into the profession of arms.

As I advanced in my behavioral science and leadership (BS&L) area of study, I took a higher level leadership class that further prepared me for commanding soldiers. In this program of study, we read leadership scenarios taken from actual battle accounts.

One such scenario was set in the thick of battle in Viet Nam. A leadership crisis had occurred in an infantry platoon. The company had seen numerous losses, as the enemy had picked off several members using tunnels to strike and then evade capture. One platoon had lost leader after leader, and then soldier after soldier. Demoralized, those who remained refused an order to advance.

The battalion commander made his way to the beleaguered platoon and, in person, commanded the remaining few men in the platoon to advance. Dispirited and sick of the death and dying, the men had lost the will to continue. After several attempts to get even one solider to take "point" and lead the platoon, the battalion commander realized he had to lay it all on the line. He then asked the platoon, "What if I take point? Will you follow me?" Then and only then did the men accede to his leadership. They followed this courageous man because he was willing to risk everything, just like he was expecting *them* to do.

And so, at first hesitatingly, the men advanced, following him. The commander was not maimed or killed, and in time, a solider stepped up and asked to take over the lead. The mission was able to be continued.

In Sun Tzu's *The Art of War*, he writes, "Regard your soldiers as your children, and they will follow you into the deepest valleys; look upon them as your own beloved sons, and they will stand by you even unto death." [1]

It is that type of sacrificial leading that flows from a deep place and can only be described as love.

The commander's willingness to lead his men so encouraged his soldiers that they recovered their confidence and willingness to fight. It is that type of sacrificial leading that flows from a deep place and can only be described as love.

---

1   "The Annotated Art of War" (Parts 10.25-26 Soldiers as Children), Retrieved May 23, 2018, from http://changingminds.org/disciplines/warfare/art_war/sun_tzu_10-4.htm.

This element of the West Point ethos had taught me to revel in "taking care" of *my* soldiers. As a young lieutenant, I knew I was fundamentally responsible for the health, welfare and training of my soldiers. We practiced and drilled as a unit. We debriefed after every exercise to improve our performance at the next exercise. When deployed for weeks at a time, I worked hard to ensure that my subordinates had adequate food, shelter and rest. As my soldiers were mostly mechanics, I took care so that they had the equipment and repair parts to do their jobs. When required, working with my chain of command, we also disciplined those soldiers who did not perform as expected or who created issues within our unit.

In time, I learned the difference between taking care of my soldiers and overstepping my role. On one occasion, we were deployed in a remote area of Germany. I decided to purchase candy to keep my soldiers motivated during their long work days. After spending my own money, there became an expectation that I do this every day. I learned from that experience that I was not there to meet their daily candy high. I was their leader and I needed to inspire and lead them to greater levels of performance rather than buy them off with sugar. In time, I came to understand that the better form of leadership required that I "love" my soldiers.

To demonstrate my "love" for leading my men and women, I needed to invest the time to get to know them and their individual situations. I spent time with them. I learned about where they were from, what they believed were their strengths and weaknesses, and to understand their perceived obstacles on the job. I inquired about their spouses and children. I learned that troubles at home had a direct impact on their attitudes and performance. In some cases, this meant getting them the help they needed (such as marriage and family counseling, financial/debt assistance and even help with drug and alcohol abuse).

As I learned more about the troubled marriages, I knew I needed to create a support system for the family members of the company. With my commander's support, we enlisted a group of wives to help us understand some of the challenges they faced when their spouses were frequently deployed. By listening and reserving judgment, we built trust with these women. We then asked for their suggestions to counter some of the stressors they experienced during their spouse's deployment. One proposal led to organizing a Company Family Day where the families of our soldiers could get to know one another and enjoy some entertainment with their spouses and the soldiers in the company. I was tasked with organizing the event. Many of our non-commissioned officers and soldiers also quickly volunteered to help. A couple of soldiers even formed a band for entertainment. The day resulted in a huge turnout of spouses and children. There were food and prizes and a new sense of camaraderie within the unit. And the company commander recognized gathering the families of his soldiers was important enough to continue. I received orders for my next assignment shortly after the event. However, that lesson in love remained as one of the most unforgettable experiences in my overseas assignment.

For a soldier, the deepest form of love appears to be forged in the crucible of the battlefield. I was recently reading *We Were Soldiers Once ... and Young*. This is an account of Ia Drang, the opening battle involving Americans in Viet Nam. It was here that 234 American soldiers fought and died. In his prologue, Lieutenant General Hal Moore referred to this battle as not just another war story, but that of a love story:

> "Another and far more transcendent love came to us unbidden on the battlefields, as it is on every battlefield in every war man has ever fought. We discovered in that depressing, hellish place, where death was our constant companion, that we loved each other.

We killed for each other, we died for each other, and we wept for each other." [2]

General Moore, USMA Class of 1945, had given and received agape love in the cauldron of combat.

Another general took this form of love into his post-military career. Major General John H. Stanford, after leaving the service, became Seattle's school superintendent as the first non-educator to hold that office. He introduced innovative ideas including a drive toward literacy. It was not unusual for General Stanford to work sixteen-hour days. In short order, the school system saw test scores rise and the performance gap between white and minority students narrowed.

His leadership philosophy is summarized in his book, *Victory in Our Schools*:

> "Most people don't expect a General to talk about love. But we talk about love all the time, because love is a key leadership principle. Love is what the famous military commanders use to inspire their troops to risk their lives in battle; it's what the most effective CEOs use to elicit maximum performance from their employees; it's what the best parents use to encourage their children to grow and learn. It's what teachers and principals use to get academic performance from their students." [3]

This old soldier brought agape love into all his endeavors. It is what all leaders *should* do and *can* do.

---

2   Moore, H. G., & Galloway, J. L. (1992). *We Were Soldiers Once ... and Young: Ia Drang: The Battle That Changed the War in Vietnam*. New York, NY: Random House., p. xviii.

3   Stanford, J., & Simons, R. (1999). *Victory in Our Schools*. New York: Bantam Books., p. 196.

All forms of love are needed to give one a well-rounded experience in life. Fortunately, I acquired and experienced all forms of love at West Point.

CULTIVATING CHARACTER

Developing all forms of love is a continual process. Great leaders understand that love is an act of the will. As Stephen Covey states, "Love is a verb" [4] — it is an action word, and one that requires intentionality.

Let's think of the various forms of love from the leadership perspective:

- *Philia* love creates the possibility for camaraderie and esprit de corps. When you show consistent regard for members of your team, department and colleagues, you invite them to show that same regard for you. You can then enjoy mutuality and respect.

- *Eros* is typically thought of in terms of romantic love, yet it can also refer to a deep passion for excellence. Leaders who inspire are absolutely passionate about their work. They love what they do and their enthusiasm spills over to their colleagues and subordinates. Do you love your work?

- *Agape* love is the highest form of love because it is both sacrificial and unconditional. It is demonstrated through selfless actions, such as taking the lead on the patrol or not taking a promotion so you have more time with your family. Agape love is demonstrated in an endless variety of forms. Yet at its zenith, it also does not have an agenda; the activity itself appears to be enough. For most of us, we are unable to consistently demonstrate agape love. Yet when we do, we may be completely unaware that we are doing so. For example, when a soldier signs up for yet another tour in the Middle East because she loves her country and her team, and she does not think

4   Covey, S. R. (2004). *The 7 Habits of Highly Effective People: Powerful Lessons in Personal Change.* New York: Free Press. p. 79-80.

twice about the sacrifice. Or when you chose to visit a friend convalescing from major surgery even though you have a demanding day at work, or you feed the homeless even though you have plenty of other things you could do with your time. When these acts of love are as natural to you as breathing, then you approach what agape love is all about.

Building your "love quotient" begins by incorporating a set of practices into your life.

1. "Seek first to understand rather than be understood,"[5] as Covey contends. This demonstrates a willingness to subjugate our own interests to those of others.

2. Assess your level of love for your work. Attempt to align your deep personal values with your work life. Connecting with your core beliefs creates the passion for the work you do. It is from passion that we are able to willingly sacrifice. If it is not financially feasible for you to align your occupation with your deep-set values, choose to connect to your passions through volunteer activities or hobbies.

3. Spend time investing in others such as teammates, employees, family and friends (for more on developing your tribe, see the Encouragement chapter). A simple means to show love for those around you is to actively listen without judgment. It is perhaps one of the most powerful means of creating bonds between people, even those who we may not understand or like, but who may have something to share or teach.

4. Accept that demonstrating love in all its forms leads to the possibility of being misunderstood, taken advantage of, and even disappointed with the outcomes. Know that setting appropriate boundaries to avoid dysfunctional behaviors and co-dependence is also an act of

---

5   Covey, S. R. (2004). *The 8th Habit*. New York, NY: Simon & Schuster, p.134.

love. Given the frailties of our human condition, I think that perhaps that is why love is also defined as an act of endless forgiveness. This implies that we must forgive others and be willing to forgive ourselves.

Love in all its dimensions is required for a full life and as a leader of others.

Love in all its dimensions is required for a full life
and as a leader of others.

ON DEATH
AND BEYOND

Given the mission of the United States Army — to preserve the peace
and security by providing for the defense of the United States — we all
knew that after graduation, we could be called into combat and that some
of us may die. We also could be harmed or perish in a training exercise
or in one of the Army's advanced training schools. A soldier's life is
dangerous.

---

*We all knew that after graduation, we could be called
into combat and that some of us may die.*

During the time I attended West Point and then in my years in the
armed forces, I was not overly concerned with this likelihood. I served
in a peace-time Army. During my active duty service, we repeatedly
rehearsed numerous scenarios to be prepared for war. However, even
when planning these exercises, I was not overly concerned that I would
die in battle.

Much that we studied in our various military science and military
art classes at West Point suggested that facing death is inevitable for
a soldier. As cadets, we mourned another cadet's passing. I also recall the
day our beloved French professor left the Academy, was assigned to the
Middle East and was blown up by a car bomb. We mourned his loss.

My husband, Eddie, would openly discuss death. His father, Jim, had narrowly escaped a premature death on the night his son graduated from high school. A splitting headache turned out to be a brain aneurysm, and an emergency flight for brain surgery saved my future father-in-law's life. Jim had a long recovery, but fortunately enjoyed many more years of working and living. Still, the uncertainty of that period in my husband's life left an indelible mark. Eddie made certain we had adequate life insurance should either of us pass away unexpectedly.

My father-in-law was a doctor in a small town in New Mexico. On occasion he might have a patient pass away. It was not uncommon in that close-knit town for Jim to attend viewings, wakes and funerals, sometimes with "little" Eddie in tow. Perhaps those early experiences with dying and its aftermath are why Eddie was more comfortable with the topic of death than most.

As an armor officer, he came home one day and asked me, "Today I learned what my life expectancy is in combat as a tank commander. Guess what it might be?" I said I was not sure. He matter-of-factly stated, "Nine seconds."

I was a bit concerned, but I believed we were in peace time and dismissed the thought.

Having completed our military obligation in 1990, we left the service and Eddie began a corporate career in paper manufacturing. I chose to run a business from home that allowed me to focus on raising our first two daughters. Having few corporate mentors in either family, and like many veterans, Eddie jumped around in his career those first few years out of the service.

Eventually we landed in Richmond, Virginia, where Eddie became an engineer for a very large, stable company. We agreed that if we were to have another child ("try for that boy"), we should do it at this time. Never having a problem getting pregnant, we were expecting within weeks of our decision.

That summer, we moved into a home intent on completing the renovations prior to the baby's arrival in October. Our days were long as we removed wall paper, patched holes and repainted each bedroom in the home.

As summer progressed, we learned that we would be having another girl. Eddie would quip, "Three daughters means three weddings — I have to be rich!" Like my dad, Eddie knew weddings can be costly. Nonetheless, we were happy and expectant of this new life. As the summer drew on, I developed sciatic nerve pain in my back and began walking with a limp. I had to cut back on some of the renovating and rest my back. Eventually, my doctor had to prescribe pain killers so I could simply get through the day.

As the birth of our baby girl drew closer, Eddie saw a flier at work looking for employees to go whitewater rafting in West Virginia. My husband loved the outdoors and the thrill of rafting. I was happy to know that he would be doing something fun with his teammates prior to our baby's birth.

I remember the night before he left on the weekend trip, he was online checking out pictures of the controlled chaos of the Class Six (VI) rapids. He called for me, joking, "Sara, come check this out. It looks like so much fun! Maybe you and the girls should come with me?" I had been in a good amount of pain that day, and I had just sat down in the easy chair and did not want to aggravate that back again. So, I never got up and looked at the pictures. I just said, "I don't think it would be a good idea for me to try to camp on the ground with this back." He agreed. But, had I not been pregnant and in pain, I would have been right there with him.

The next morning, I drove him to work so he could catch a ride after work with his co-workers. It was a thirty-minute trip. The sun was shining, and I remember his muscular arms holding the steering wheel. I can't forget looking over, admiring my strong, handsome husband. When I dropped him off, we said goodbye with a kiss.

I went about that day and the next running errands. I was happy thinking that he was enjoying himself. The baby would be here soon, and our lives would become even more complicated. I came home, made dinner, and was easing myself into a chair when there was a knock at the front door. When I opened it, there was a law enforcement officer and two of the pastors from my church. Confused, I thought there may have been a problem at church, but I thought, "Do they really need a police officer to relay this kind of information?"

I asked them to sit down, but the officer refused a chair. He was intent on determining if I was James Edwin Gaba Jr.'s wife. I replied, "Yes. I am." He then asked if Eddie was on a rafting trip in West Virginia. I also confirmed this fact. Then he said, "He has been lost."

Not yet taking in what he said, I assumed he meant lost in the woods. And, if that was the case, I was not concerned because Eddie was an Airborne Ranger and he would be eating berries and surviving until someone found him or, more likely, he made his way back to civilization. The officer's next words filled me with grave concern. "No ma'am, he was lost in the water." He handed me a piece a paper and continued, "You need to call this number and speak with a forest ranger from that area."

I knew then that this was not going to be good news. Upon reaching the ranger, I soon learned that my husband had drowned. I would later discover that the rafting company had overbooked for the opening day of the season. They hired an inexperienced rafting guide who was unfamiliar with the river. The guide hit a rock, aptly named "shipwreck" rock. Eddie

Eddie, Sara (pregnant), Larisa, and Gwenyth, 1997

was thrown out of the boat; immediately pummeled by the rapids, pulled under and pinned under a rock, he drowned. Only after the dam had been stopped and the waters receded for four hours was it possible for divers to enter those waters and retrieve his body.

So, at eight-and-half months pregnant, I was widowed. The most difficult thing I then had to do was to try to explain to our daughters (who were just nine and seven years of age) that the dad they adored was gone. Next, I had to call his parents and tell them that their only son was dead. I still recall the piercing cries from his mother, Paula. Cries that penetrated my soul as we both recognized the finality of Eddie's passing.

The organizer of the rafting trip had arranged for my pastors to be there for me when this heart-wrenching news was given. I am so thankful that Brian Rich made this effort. My pastors prayed with me and arranged for a neighbor to stay with me that night.

I was very fortunate to be a part of a wonderful and supportive church. The men of the church stepped up and helped by coordinating all the family members who were traveling from out of town. These kind men shielded me from many phone calls and further emotional pain. Eddie's company and those fellow rafters offered encouragement and support. My brother Lee and my sister-in-law Kanda came for my baby's birth. Over the years, other family members including my brothers Sam and Dave and my sister-in-law Taryn and my brother-in-law Scott all traveled out to Virginia to support me and my girls. My neighbors were also incredibly supportive. They hosted my family during the funeral and then provided meals for months. They even completed my new baby Joye's room prior to her birth. The men of the West Point Society of Richmond were instrumental in my efforts to restart my career. I am indebted to all these wonderful people to this day.

Despite all of this support, there are just some things a widow must do all on her own. It was at this tragic, lonely time that I became truly grateful for my West Point experience. To do what needed to be done, I simply began to give myself orders. I remember saying, "Lieutenant, you will go to the funeral home and pick out a casket today." "Tomorrow, you will choose what suit and tie he should wear in the casket." "You will call the insurance company and find out what is needed to get his life insurance." I believe I choose to call myself a lieutenant in deference to all the less-than-enjoyable jobs dumped on junior officers.

It was at this tragic, lonely time that I became truly grateful for my West Point experience. To do what needed to be done, I simply began to give myself orders.

The discipline I learned first as Plebe and then as a young officer aided me in handling the enormity of this personal tragedy. In time, I would hire and fire a financial planner, sell an old car and buy a new one, and sell and buy a new home. I formed a support group with other widows and served as a board member for a camp for grieving children. And I volunteered to mentor grieving families after 9/11. I restarted my career in a corporate role and later developed my own consulting practice. In time, I expanded my business and added contractors to help me accomplish the projects I was accumulating. I completed an advanced degree while working full time. And, I learned how to parent alone – this was by far the most difficult of all my leadership roles.

My parents each passed way within a few years of Eddie's death. My in-laws lived thousands of miles away in Colorado. I had hoped for their emotional support. Yet, the distance combined with their own grief made them less available emotionally to me than I had hoped.

I recall that my father-in-law would call and tell me about a friend or co-worker who had inquired about how the girls and I were doing. He would then relate that he replied to his friend, "Well, Sara is a West Pointer. She is a wonderful mother and is doing a great job with those girls."

At the time, I wanted more emotional support from Jim and some recognition of how difficult it was for me. Did he know how challenging it was to work full-time and raise three girls? Yet, "She is a West Pointer" was a common phrase that he often used to describe me at the time.

Now that I have had years to heal from the trauma of having Eddie ripped from my life, I know that what Jim was saying was absolutely true. Even if my father-in-law did not fully comprehend the extent of the challenges I endured being a West Point woman, he understood what it meant to have graduated from the Academy. The discipline and challenge of being a woman at West Point had prepared me for the greatest leadership challenge of my life — learning to raise three strong daughters, and to live and thrive even after this heart-wrenching loss.

You might ask, how did you persevere? By employing all the essence of the West Point experience and then putting one foot in front of the other, over and over, again and again. Mine is a journey I hope may encourage you, even in the face of incredible personal tragedy.

CULTIVATING CHARACTER

For me, developing West Point character is never more powerful than when we are faced with a heart-crushing blow like the death of a loved one. While I am only briefly noting four suggestions on how to overcome the loss of a loved one, I hope that the West Point axioms that I employed during my own time of loss help to support you:

- **Take care of the important things first.**

   Doing the right thing. I chose to take care of the important things first even if at the time it was difficult to do so. Initially, I sought counseling individually for myself and then found a counselor for my daughters. Those first appointments were incredibly painful as I was distraught, scared and unable to make sense of the loss. For my two daughters, they did not necessarily enjoy going to a counselor and talking about the loss of their dad; however, I knew this was healthy and needed to be done. Later we had family therapy and it was never easy to open the wounds and manage the pain. I still recall my therapist admonishing me to take care of myself, saying, "Your children will only do as well as you do." So we kept going and then we would take a break and return many times.

   And although one does not want to think about one's own passing after losing a spouse, as I was eight-and-a-half months pregnant, I knew that I needed to make arrangements for my two young daughters should anything go wrong during the delivery of my baby. So not long after the funeral, I made efforts to secure care for my daughters in the event they were to lose their only other parent (me). I drew up a new will that stipulated their guardians, secured medical insurance, and eventually increased my life insurance benefits.

- **Develop an emotional and professional network of positive, supportive connections.**

  Camaraderie and encouragement. Early on I joined a widow support group. Later I formed one of my own as I met several women who lived nearby and had also recently lost a spouse. Working through our grief together provided both for healing and a source of friendship. As time progressed, I developed a network of several positive, supportive people. One of my "peeps" is Doris Harkness Birdsong. Widowed only about eighteen months before I was, Doris was on the same rafting trip when Eddie died. She freely offered her time to encourage me even as she ran a business and raised her *own* two boys. Doris became an enduring source of strength to me and my daughters. "Aunt Doris" is as generous in her love as in her wit, and I thank God for her.

  There were also professional associations I developed, many of which are with West Pointers, and some of which I have mentioned in this book. These relationships led to meeting graduates from all the service academies and then, from there, linkages to several business contacts. These connections catapulted my career.

- **Rely on solid reasoning to make major decisions after a loss. Delay a decision if you are overwhelmed.**

  Problem solving. Anyone who has suddenly become a single parent knows that there are a plethora of issues the surviving parent will encounter. These range from managing the emotional needs for each grieving child, to making wise financial decisions, to securing adequate childcare, to coordinating carpooling to doctor appointments and extracurricular activities, to getting household chores and yard work completed. And we must do this all while working full time to support our family. Leveraging practical problem-solving skills and delegating some efforts to others was how I tackled each issue that arose. As I learned in that fifty-meter pool at West Point,

there are myriad ways to solve a series of problems, but we all have to
come up for air now and again.

• **Seek help when overwhelmed.**

Humility. We all have our limits. I felt woefully inadequate in some
of the roles that Eddie had always assumed. For example, he had
made our financial investment decisions and suddenly I needed to
make decisions on more money than I had ever handled. He had
taken care of our cars and spoken with the mechanics to ensure
a repair was actually necessary. Now I was thrust into these situations. Often, the combination of feeling incompetent while dealing
with my own grief left me overwhelmed. So, I sought out help.
I sought out experts in areas I knew I was incompetent. As time
passed, I solicited help with re-starting my career, securing a quality
babysitter, and locating a quality financial planner. And I continued
to see a counselor especially after starting a demanding new job
and when my daughters entered those difficult teenage years. My
advice to you is to recognize that it is okay to admit that you cannot
do it all.

Understanding Grief

In life, we will all face many losses. We may experience: a divorce,
a health crisis, a loved one with alcohol or drug dependency, a loved one
who is incarcerated, job loss, financial disaster, or foreclosure. The list can
go on and on.

As a leader, we will certainly have employees who will face personal
tragedy and loss as well. Being equipped to lead them through these
times is essential to the role of a leader.

An incredibly helpful tool for understanding loss is the Kübler-Ross Grief Cycle.[1] Through my own grief, I came to understand that all loss is a process, one with a beginning and an end. There are predictable emotions that one can experience as we assimilate the loss into a new reality.

Kübler-Ross Grief Cycle

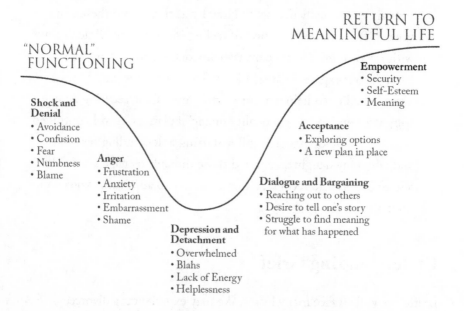

RETURN TO
MEANINGFUL LIFE

"NORMAL"
FUNCTIONING

Shock and Denial
• Avoidance
• Confusion
• Fear
• Numbness
• Blame

Anger
• Frustration
• Anxiety
• Irritation
• Embarrassment
• Shame

Depression and Detachment
• Overwhelmed
• Blahs
• Lack of Energy
• Helplessness

Dialogue and Bargaining
• Reaching out to others
• Desire to tell one's story
• Struggle to find meaning for what has happened

Acceptance
• Exploring options
• A new plan in place

Empowerment
• Security
• Self-Esteem
• Meaning

Initially one experiences shock, then denial, and this is followed by anger and depression. In time, you begin to bargain and experiment with how to deal with the loss. Eventually, you learn to accept the loss and then re-establish yourself in a new form of identity. One benefit of this model is that it provides labels for the emotions that we can experience. By

---

1   The Kübler-Ross Grief Cycle. (n.d.). Retrieved May 22, 2018, from http://changingminds.org/disciplines/change_management/kubler_ross/kubler_ross.htm.

naming those emotions as they occur, we can diminish their intensity to some extent.

Although the model suggests that the process is linear, one step following another, the grief journey in my experience is more like an endless spiral staircase. It is one where you experience these emotions repeatedly, and sometimes several of them at once. As time passes, the depth of the pain recedes and is replaced with more positive memories. I now recall the funny, delightful memories, sometimes above the painful years of mourning. But there are still moments and events that reopen old wounds.

For years, Valentine's Day was especially painful for me. And then significant occasions, such as when each daughter graduated from high school, or when Larisa and then Gwenyth got married. It is then when we all hurt once more as we grieve the fact that their dad was not there for these major life events.

A widower friend of mine summed up what it feels like to lose a spouse:

> "It is like losing a leg and then trying to ride a bike again. You learn how to pedal with the other leg. You have to work harder at pedaling and eventually you learn how to do it. But you will always miss that leg."

I do know that I had very few regrets in my marriage to Eddie Gaba. It would have saddened me if I were as other widows I have met who had unfinished business with their husbands. They experienced remorse for an argument before losing their spouse or for not saying , "I love you," enough.

I did not have those same regrets with Eddie. We knew we loved each other unconditionally, and although I miss him even today, I also know that I will also see him again. Eddie and I shared a deep Christian faith that assures us of eternal life. And I knew from the moment I heard he was gone that he was in fact alive and in heaven. I grieve but not as one without hope because I am certain that we will be reunited in time.

The day to day realities of life are very real and were, at times, more than I thought I could bear. What I discovered was that I could navigate the travails of life when I relied on the principles imbued through those challenging years at the Academy. By applying those lessons, it made my widow's walk no less painful, but in every way, possible. It is my hope that you also will find these truths enabling should you face a dark night of the soul.

THE DREAM

I found myself in cold, turbulent water and I was struggling to keep my head above the waves. I looked around, searching for the shore even as the waves of water hit my face again and again. Then I saw them, a few others near me, struggling in the water along with me. A few were women and others were men. I lifted my arms over and over, and the water filled my eyes and mouth. As I stroked, in the distance I saw the shore. Even as I doubled my effort, I did not seem to make progress. I began to get discouraged. As I treaded water and rested a bit, I saw that some of the swimmers had turned back and were swimming with the current and making good progress. I shouted at them to hang with the few of us who endured, but they did not hear me and paddled away.

Yet I realized my direction was set and I was not destined for an easier swim. As I resumed my paddling again, the swimming became even more difficult. I was being pulled down. It was then I realized I was dragging a pouch shackled to my leg. The sack pulled me down, and I worked harder to stay afloat. First my arms began to spasm and then my legs began to atrophy. Each stroke was painful, and my legs began to give out. I worked harder to stay afloat, and I cried out in pain. I stopped swimming and turned on my back and it was then, that I saw them away in the distance.

Over on the far shore of the river, a few women and men were emerging from the water. They appeared to be looking at me, waiting for me. They each carried a torch, and as the day waned, they lit their lanterns so we could see them. What's that they were saying? They were shouting at me. They were telling me that I would make it and to keep going. They began to point out how to make the best direct line to the shore. I heard them yell directions about how to avoid whirlpools of current that might take me away from them. And as they lifted their voices and shouted out their wisdom, I got a second wind and my strength was renewed.

As the shore came closer, I realized there was a limit to the pain. It was only a matter of time. I became convinced that I could hang in there a bit longer. Slowly, the faces of my troupe became clear; the golden light from their torches lit their faces. My compatriots were patiently watching and waiting for me. As I drew near, their cheers got louder and more distinct. I kept my eyes on my supporters and forced myself to relax and keep the momentum going. Slowly, and bit by bit, that beach drew closer and closer. And then suddenly, my feet were underneath me. And I could feel rocks and then pebbles and then sand mixed with mud oozing between my toes.

I stood upright and emerged from the cold waters and began to run. I ran up the shore and onto the beaten mud path. There were rocks and mud and I slipped and hit a knee. I groaned but got back up and noticed that those people from the shore were now in front of me; some of them were also beside me and behind me. And as we labored, I could hear them breathe. Like me, they were also struggling. I reached out and helped one of the other cohorts as we ascended a hill. I encouraged another as we adapted to the rough terrain and changing landscape. And then our breathing became unified and calming. As we marched up and over a hill and kept up our pace, I noticed that we were once again climbing.

And then our breathing became unified and calming. As we marched up and over a hill and kept up our pace, I noticed that we were once again climbing.

A magnificent mountain lay in front of us and our troupe was winding around its sphere. We were always ascending and climbing higher. At each turn, the struggle and pain was real, yet we were one in our desire to reach the pinnacle. We no longer complained; we began to enjoy the rhythmic and mutual effort of our dance. We sang songs and beat on our chests, all the while continuing our forward momentum. "Onward, upward, together we will never return." We chanted as we were transported to an ethereal fleeting moment of serenity.

As we turned once more to the right, we spied the summit. And there on the final ascent were several young women and men. They appeared to be watching us, even expecting us. They were calling out to us, but we could not hear what they were saying. As we drew closer, I saw the eagerness in their faces. The longing to make us proud. They reminded me of myself when I first found myself in the water. They were wondering if they could do what we had done. I smiled as I thought about how difficult it had been when I first experienced the cold and the current and the drag of the weight bound to my leg. Those earnest faces were seeking to know how we had done what we did, and as we approached them, they were silent, respectful.

They extended their arms toward us, as though they were asking us to hand off to them our knowledge of the swim, the climb and the struggle. In so knowing, they might avoid some of the dangers and swirling pools that we had encountered. And that is when I noticed that our mountain top was just the beginning of their journey.

In front of the youth before us lay a deep mountain river and another mountain peak on the other side. I realized that it was their turn now to enter the cold and turbulent water and begin to swim against the current in their journey. I could now understand the voices of the assemblage that had encouraged me. They were asking us to begin the adventure with them. To grip hands with them as they entered the cold, uncertain waters.

To share with them how we had overcome the struggles and the pain of our journey.

They were asking us to begin the adventure with them. To grip hands with them as they entered the cold, uncertain waters. To share with them how we had overcome the struggles and the pain of our journey.

And as evening set in, we lit our torches and began to tell our stories. Each of us reached out our hands to one of the young men and women who we had shared this experience with, and we clasped them tightly. We told them of the struggle, the pain and the ability to overcome. Our stories of both failure and fortune, of life and loss would embolden them in the struggles they would certainly face in life.

When we were long gone and merely ghosts in their memories, I was hoping that our words would continue to compel them to even greater heights. I wanted them to reach the higher ground that we had never endeavored to attempt. And then, after these youths had given their all to their climb, there waiting for them would be their progeny holding out their arms to embrace them. Thus, the circle continues with each one, gripping hands with those who came before and who would reach back to those who come behind. We expect the older to engage and enable the younger. This striving that we seek is worth all the effort and the glorious mountain tops that await each successive generation.

I then awoke and then realized that this vision was the essence of the West Point experience. And so, I encourage you, dear reader, to take the plunge and learn from the journey of a West Point Woman. Then may you climb with confidence.

As all members of the Long Gray Line sing in unison … Grip Hands!

AFTERWORD

There are now more than 5,500 women who have graduated from one of the most challenging academic institutions in our country. West Point women have been tested physically, intellectually, emotionally and spiritually while still facing skepticism and shattering preconceptions. All of us have unique stories of our own.

We women are now working in all walks of life. We serve on the front lines of Afghanistan and Iraq, leading large Army units into combat. We contribute to society as corporate leaders, civil servants, scientists and medical doctors. We are running for elected office, and many of us are raising the next generation of leaders. We are your colleague, neighbor, sister, daughter or wife. What you must remember is that you must never underestimate a West Point woman. She is destined to surprise you.

It is as a mother, and not a coach or consultant that I wish to end this book. Let us return for a moment to that scene I described in the Prologue. After delivering the speech to that women's leadership group, I fielded several questions from the audience. One that I was keen to answer concerned my daughters, whom I had only mentioned briefly. A woman inquired as to what they had done and were doing with their lives.

I can say without hesitation that Larisa, Gwenyth and Joye are all strong, independent women. And although they did not pursue attending West Point or a military career, they each chose educations and career paths that reflected their individual passions and talents.

Joye, Gwenyth, Sara and Larisa at Gwenyth and Leo Thome's wedding. Oct 7, 2017

Larisa initially chose teaching as a career. However, because of her deep desire to alleviate the suffering of others, she recently became a registered nurse. Gwenyth is an artist and filmmaker, and just completed her third master's degree to prepare her to create artistically beautiful films that serve a higher purpose. Joye chose commercial interior design and recently graduated college and landed a job with a commercial architectural firm. Even though she never had the opportunity to know her dad, she inherited his keen eye and sense of order.

Despite the pain of losing their dad and having a less-than-perfect mom, they have all turned out to be smart, hardworking and compassionate women. My progeny had in fact, caught the very essence of the West Point character – the very precepts I had patterned my entire adult life upon. And for that, I am truly thankful.

Although my children did not attend the Academy, I would encourage any young person who is seeking a challenge, has a desire to make difference and who wants a life-changing college experience, to apply to a service academy. However, I would not be a West Pointer if I failed to not recommend my alma mater as the best of those institutions to consider.

I also want to encourage all West Point women to share their own unique stories. In this way, others can benefit from your personal journey. Record for your progeny and the world the evolving story of our rock-bound highland home. And then all may benefit from the passing of the Long Gray Line.

 West Point Women: Post your stories and photos in Sara's closed Facebook group, "West Point Woman."

https://www.facebook.com/groups/westpointwoman

ACKNOWLEDGMENTS

There are so many important people in my life who contributed to making this book a reality.

- My wonderful, loving husband, Reed M. Potecha, for his encouragement and enthusiasm while I was writing this book and for reading endless revisions and adding great insight as a fellow veteran. This Navy guy has proven himself worthy, over and over again.

- My daughters, Larisa, Gwenyth, and Joye, who inspired and encouraged me to tell my story.

- My brother Dave, who has spent hours reading the manuscript and supplying much-needed improvements. Thanks, Bro!

- My editorial board: Elizabeth Jeffries, Lesa Nichols, Barney Forsythe, Susan Kellett Forsythe, Gail O'Sullivan Dwyer, Karen McCombs, and Kathy Schmidt Loper. Thank you for your input on how to improve my manuscript. And a special thanks to Elizabeth Jeffries for suggesting the title *West Point Woman*.

- My gifted book coach, Cathy Fyock, for encouraging me every step of the way with the writing of the initial draft. Thank you for also connecting me with organizations to hire me as a speaker.

- My new publisher and all the staff at Emerge. Thank you Christian Ophus, Richard Robertson, Julie Werner and Megan Ryan. Your efforts have made this second edition better than I ever imagined.

- To my classmates who contributed photos: Jan Tiede Swicord and Lorraine Lesieur. And my cross-country teammate, Harlene Nelson Coutteau, USMA '82, who supplied photos of our winning XC team.

- I am also indebted to Alicia Mauldin-Ware, USMA archives curator, for many of the historic photos in this book.

- My West Point classmates and Academy friends who have contributed to the tales contained within. Thank you for those memories.

- And, always, I thank my Heavenly Father who has blessed me with many abilities and who, despite all my shortcomings, is incredibly patient, showering me with unconditional love.

GLOSSARY OF CADET VOCABULARY

Butterbar – A second lieutenant.

Beast – Cadet Basic Training (CBT).

Boodle – Cake, candy, ice cream, etc.

Butt – The remains of anything, as the butt of a month. When called to recite the days for upperclassmen, Plebes would need to know how many days there were until important events like the Army-Navy game, graduation, etc. As an example, a Fourth Classman would state, "Sir (or Ma'am), there are thirty-six and butt days until Army beats the hell out of Navy!"

CFT – Cadet Field Training, summer training for Third Classmen, held at Camp Buckner.

Civvies – Civilian clothes.

COM – The Commandant of Cadets.

Cow – A member of the Second Class.

Source: United States Military Academy. (1979). A glossary of cadet slang. In Bugle notes '79 (71st Volume, pp. 312-319). West Point, NY: United States Military Academy.

Crab – One who attends the Naval Academy, also squid.

D – Deficient, below average in academics.

Firstie – A member of the First Class.

Fried Egg – Insignia of the USMA worn on the headpiece.

Goat – A cadet who graduates at the bottom of his or her class.

Green Girl – The comforter at the end of a cadet's bed. As women, we referred to them as "Olive Gentlemen."

Gross – Blundering, dull.

Hop – A cadet dance.

Juice – Electrical Engineering.

Max – A complete success, such as "I maxed that chemistry exam."

O.C. – Office in Charge.

O.D. – Olive Drab.

Odin – A Norwegian god to whom cadets appeal for rain before parades, inspections, etc.

P – a professor or instructor.

PDA – Public display of affection. Cadets were to refrain from this.

Ping – To move out at 120 steps per minute. Plebes during my time at West Point were required to ping from place to place.

Plebe – A cadet of the Fourth Class, a Freshman.

Plebe Bible – *Bugle Notes*, the handbook of the Corps of Cadets.

Police – To throw away or to clean up, as in "police the grounds" around a building.

Poop – Information to be memorized.

Poop Deck – The balcony in the dining hall where the O.C. eats and from which orders are published.

Pop off – Sound off in a military manner.

Post – Short for "take your post." To go about your business.

P-rade – A parade.

Pro – Proficient; above passing in studies or looks.

Rack – Cadet bed, also a sack.

Recognize – To place a Fourth Class cadet on upper-class status.

Roger – I understand.

SAMI – Saturday morning inspection.

Slug – A special punishment for a serious offense. Or, to impose a special punishment on a cadet.

Solids – Engineering mechanics.

Sound off – To shout.

S.O.P. – Standing operating procedure.

Spaz – To function improperly, typically referring to New Cadets and Plebes who performed something incorrectly.

SOSH – Social sciences. Referring to the infamous SOSH paper.

STAP – Summer Training Academic Program, or summer school.

Star man – an academically distinguished cadet who is given stars to wear on uniforms.

Striper – Rank of cadet captain or above. Sometimes called a "striper dog."

Supe – Superintendent.

TAC – An officer in the Department of Tactics. One TAC is assigned to each of the thirty-six cadet companies.

T.E.E. – Term end exam, finals.

Turnback – A re-admitted cadet.

Woo-Poo-U – West Point, also Woops.

Wopper, W.O.P.R. – Written Oral Partial Review, an oral exam.

Writ – A written recitation, an examination.

Yearling – A member of the Third Class, also a Yuk.

Zoomie – One who attends the Air Force Academy.

OTHER BOOKS BY
"WEST POINT WOMEN"

Barkalow, C., & Raab, A. (1992). *In the Men's House: An Inside Account of Life in the Army by One of West Point's First Female Graduates*. New York, NY: Berkley Books.

Beaudean, J. (2011). *Whatever the Cost: One Woman's Battle to Find Peace with Her Body*. Bolivar, MO: Quiet Waters Publications.

Furdek, C. (2017). *Locked-In: A Soldier and Civilian's Struggle with Invisible Wounds*. Louisville, KY: Arden Piper Publishing.

McAleer, D. M. (2010). *Porcelain on Steel: Women of West Point's Long Gray Line*. Jacksonville, Florida: Fortis Publishing.

O'Sullivan, D. G. (2011). *Tough as Nails: One Woman's Journey through West Point*. New York: Midpoint Trade Books.

BIBLIOGRAPHY

About West Point – Mission. (n.d.). Retrieved March 15, 2018, from https://www.usma.edu/about/SitePages/Mission.aspx.

American Rhetoric Movie Speeches. (n.d.) Patton 1970. Retrieved March 26, 2018, from http://www.americanrhetoric.com/MovieSpeeches/moviespeechpatton3rdarmyaddress.html.

Bob Sherwin, Contributor. (2014, January 24). "Why Women are More Effective Leaders Than Men." Retrieved March 15, 2018, from http://www.businessinsider.com/study-women-are-better-leaders-2014-1.

Cloud, H., & Newbern, G. (2017). *The Power of the Other: The Startling Effect Other People Have on You, from the Boardroom to the Bedroom and Beyond — and What to Do About It.*

Covey, S. R. (2004). *The 7 Habits of Highly Effective People: Powerful Lessons in Personal Change.* New York: Free Press.

Covey, S. R. (2004). *The 8th Habit.* New York, NY: Simon & Schuster.

Cowley, R., & Guinzburg, T. (2002). *West Point: Two Centuries of Honor and Tradition.* Canada: Warner Books: An AOL Time Warner Company.

Definition of ETHOS. (n.d.). Retrieved February 22, 2018, from https://www.merriam-webster.com/dictionary/ethos.

Graham, D. R. (n.d.). "Bugle Notes: Learn This!" Retrieved March 17, 2018, from http://www.west-point.org/academy/malo-wa/inspirations/buglenotes.html.

Hickman, K. (2008, November 14). The GI General: General Omar Bradley. Retrieved March 17, 2018, from https://www.thoughtco.com/world-war-ii-general-omar-bradley-2360152.

Janda, L. (2002). *Stronger than Custom: West Point and the Admission of Women*. Westport, CT: Praeger.

Kegan, R., & Lahey, L. L., & K. (2009). *Immunity to Change: How to Overcome It and Unlock the Potential in Yourself and Your Organization*. Boston: Harvard Business Review Press.

Lencioni, P. (2012). *The Advantage: Why Organizational Health Trumps Everything Else in Business.*

Lone Sentry (n.d.) Retrieved March 17, 2018, from http://www.lonesentry.com/gi_stories_booklets/8thinfantry/index.html.

Metal Gear Wiki. (n.d.) Tap Code. Retrieved March 26, 2018, from http://metalgear.wikia.com/wiki/Tap_code

Moore, H. G., & Galloway, J. L. (1992). *We Were Soldiers Once ... and Young: Ia Drang: The Battle that Changed the War in Vietnam*. New York, NY: Random House.

Sanders, T. (2002). *Love is the Killer App: How to Win Business and Influence Friends*. New York: Random House Audio.

Schwarz, R. (2017, October 30). "The 'Sandwich Approach' Undermines Your Feedback." Retrieved June 14, 2018, from https://hbr.org/2013/04/the-sandwich-approach-undermin.

Stanford, J., & Simons, R. (1999). *Victory in Our Schools*. New York: Bantam Books.

"Stress relief from laughter? It's no joke." (2016, April 21). Retrieved May 9, 2018, from https://www.mayoclinic.org/healthy-lifestyle/ stress-management/in-depth/stress-relief/art-20044456.

"West Point in the Making of America." (n.d.). Retrieved June 28, 2018, from http://americanhistory.si.edu/westpoint/history_2.html.

"The Annotated Art of War" (Parts 10.25-26 Soldiers as Children), Retrieved May 23, 2018, from http://changingminds.org/disciplines/ warfare/art_war/sun_tzu_10-4.htm.

The Kübler-Ross Grief Cycle. (n.d.). Retrieved May 22, 2018, from http://changingminds.org/disciplines/change_management/kubler_ross/ kubler_ross.htm.

United States Military Academy. (1979). A glossary of cadet slang. In *Bugle Notes* '79 (71st Volume). West Point, NY: United States Military Academy.

ABOUT THE
AUTHOR

Sara Potecha is an accomplished author,
speaker, and consultant who has led
cultural transformation initiatives for
Fortune 100 and 200 companies and
coached hundreds of leaders to higher
levels of performance. A masterful storyteller, Sara is a sought-after speaker on
such topics as women in leadership,
veteran employment, emotional intelligence, change, resilience, and leadership

competency. Sara holds a bachelor's degree in general engineering from
the United States Military Academy (USMA) and a master's degree in
organizational management and development from Fielding Graduate
University.

In *West Point Woman*, Sara delves into fundamental leadership precepts
taught to her at a tender age at the Academy and applied in all her varied
roles. She captures her audiences through powerful, and often hilarious,
personal stories while teaching her readers how to cultivate the character within.

When Sara is not writing or speaking, you will find her outdoors — on
a bike, a hike, or boating with her husband and family. *West Point Woman*
is her first non-fiction business book.

A portion of the proceeds from **West Point Woman** *book sales will be donated to veteran services organizations that support women veterans and wounded warriors and their families.*

Connecting with *West Point Woman*

Get a conversation started with Sara at:

📞 (201) 416-9791

✉ sara@sarapotecha.com

🌐 www.WestPointWoman.com or

🌐 www.SaraPotecha.com

Go Beyond the Book

<u>Hire Sara to:</u>

- Deliver an inspiring and masterfully told keynote based on the West Point Woman leadership principles, or on a variety of other timely topics

- Provide exceptional yet practical training on a key subjects such as leading major change, emotional intelligence, micro-habits and so much more

- Drive deep levels of transformation in your organization with her proven West Point Woman Mastermind methodology (see next page)

- Coach you or your high-potential employees to higher levels of performance

WANT MORE *WEST POINT WOMAN?*

Coming in early 2021

Given the realities of our COVID world, several of our clients are struggling with equipping their leadership with the tools to meet the ongoing high degree of uncertainty, change and turmoil. In response, my team has developed a proven, effective and efficient process for leadership development that results in leaders who effectively lead through adversity.

Imagine a truly innovative **Leadership Development Experience** that could transform your organization. Based on my 2nd Edition of *West Point Woman: How Character is Created and Leadership is Learned*, we developed an accompanying Mastermind methodology that teaches proven leadership tenets to navigate chaotic and turbulent times. The Mastermind can be done in-person or virtually. It will drive lasting organizational transformation by working with key leaders who delve more deeply into the leadership skills and behaviors needed in a COVID world. **By investing in your leaders through this innovative program, you will raise the 'leadership lid' of the entire organization.**

Your leaders will dialogue about key leadership values in times of crisis. For example, they will discuss how **doing the right thing** is foundational in times of difficulty, how **camaraderie and feedback** impacts performance, and even how **humor** aids in navigating difficulty and major change. The inculcation of leaders living these values is accomplished through four key components of my program: **leadership training through storytelling, focused dialogue, honest reflection, and personal accountability.**

To get updates on when this exciting and novel methodology will be available for purchase, go to:

www.westpointwoman.com/mm